Laser Physics and Spectroscopy

Laser Physics and Spectroscopy

Pradip Narayan Ghosh
Department of Physics
University of Calcutta

CRC Press
Taylor & Francis Group
Boca Raton London New York

CRC Press is an imprint of the
Taylor & Francis Group, an **informa** business

CRC Press
Taylor & Francis Group
6000 Broken Sound Parkway NW, Suite 300
Boca Raton, FL 33487-2732

First issued in paperback 2023

CRC Press is an imprint of the Taylor & Francis Group, an informa business

ISBN-13: 978-1-138-58827-1 (hbk)
ISBN-13: 978-1-03-265337-2 (pbk)
ISBN-13: 978-0-429-49239-6 (ebk)

DOI: 10.1201/9780429492396

Publisher's Note
The publisher has gone to great lengths to ensure the quality of this reprint but points out that some imperfections in the original copies may be apparent.

Library of Congress Cataloging in Publication Data
A catalog record has been requested

Visit the Taylor & Francis Web site at
http://www.taylorandfrancis.com

and the CRC Press Web site at
http://www.crcpress.com

LEVANT

Dedicated to my Mother

Preface

The subject-area of Laser Physics and Laser Spectroscopy has developed a great deal over the last few decades. Much new information is added in the field over this period. Most of the books published so far are at very advanced level meant primarily for researchers. They are not easily comprehensible for undergraduate students. Many of these books are on Quantum Optics or biased to Optics. Usually, they do not emphasize theoretical and mathematical treatment at a level appropriate for UG and PG core courses. But theoretical treatment and phenomenological interpretations are both essence of Physics learning process.

For the purpose of teaching Laser Physics we always have to consult a number of books on Laser Physics, since it is difficult to find one or two books that could cover the syllabus. The students have to depend on the class lecture materials. So there is a dire need for a text book for both UG and PG students that can be used by Indian and foreign students. The book must start from an elementary level with the description of physical phenomena. It should also emphasize the required mathematical derivations. The book must be illustrative and should have both worked out exercises and problem assignments. My experience of post graduate teaching as well as my involvement with undergraduate examinations of Calcutta University motivated me to write the book. My aim has been to write a book such that no other book available in the market is comparable with this book in terms of the content, outlook and approach.

For the sake of completeness we have added some sections in the book that cannot be taught in the fixed time schedule. The book is a text book on the broad subject of Laser Physics and Spectroscopy. But it may also be considered as a course on Introductory Quantum Optics. However, the direction is not towards Quantum Optics, so some of the topics normally found in advanced books on Quantum Optics are not included here.

The book presumes that students or general readers are familiar with the basic concepts of Quantum Mechanics and Classical Electrodynamics. We have aimed to make it a complete text so that no extensive help is necessary from other books or references. We have enlisted the related text books on the subject for further reading. Since the book is a text for students we have avoided references to publications in journals except in a few cases. Considering the fact that basic principles of operation of laser are taught in most universities as the basic level core courses and that the students should also have a familiarity with the advantages and limitations of laser radiation, we have introduced a chapter with details of basic principles and properties of lasers. This will be helpful to the undergraduate students for the elementary course on lasers, as well as to the post graduate students in many universities or institutes.

We begin the book with an Introduction to Lasers in Chapter 1. The basic principles of operation of laser and its different essential components are discussed. Stimulated emission, population inversion and feedback oscillation are introduced. Operational principles and actions of different types of lasers with gaseous, liquid and solid state active materials are described. Recent developments on quantum well lasers are also presented. We have discussed the basic properties of lasers like temporal and spatial coherence and directionality and monochromaticity. The shape of the laser beam and its propagation are also discussed in detail. We have also considered the mode pattern in the laser resonators.We have discussed the pulsed operation of lasers and have explained Q switching and mode locking behavior. We have also presented the non-linear effects in laser physics, particularly the Second Harmonic Generation (SHG) and Optical Parametric Oscillation (OPO). Asimple Rate Equation theory is presented to explain the threshold behavior of laser operation.

Chapters 2-4 describe the atom field interaction. In Chapter 2 we consider the semi-classical interaction of radiation with atomic systems for both weak and strong fields and introduce the concept of Rabi oscillations. Chapter 3 introduces the density matrix of statistical ensembles and develops the density matrix equations by including the decay phenomena and by using the Rotating Wave Approximation (RWA). For solutions of the optical Bloch equations we have considered the Bloch vector model. The cases of moving atoms have also been discussed. In chapter 4 we consider the moving atoms in a standing wave and introduce the concept of Lamb dip, crossover resonance dip, optical pumping and closed transitions. Saturation absorption spectroscopy experiments are described with Rubidium atom energy levels as example. Atomic energy levels and the hyperfine structure of the levels are described in detail.

In Chapter 5 we present the semi classical theory of laser action. Starting from the Maxwell's equations of classical electrodynamics we develop the Lamb self-consistency equations. From the solution of the equations we obtain the intensity equation in terms of linear net-gain and self-saturation coefficients and also the frequency equations in terms of frequency pushing and pulling coefficients. We have used rate equation approximation for the calculation of polarization of a single mode and have introduced the concept of hole burning. We have also considered the cases of two-mode and multimode operations.

Chapters 6-7 describe the quantum nature of radiation and the quantum theory of atom field interaction. In Chapter 6,we develop the Hamiltonian of the electromagnetic field confined in a cavity in terms of the classical harmonic oscillator Hamiltonian. The quantization of the Hamiltonian is described and number states and photon concept are introduced. Instead of using quantum field theoretic concept, we consider the direct process of quantization from the Maxwell equations. We also consider the multimode radiation field and have defined coherent states. In Chapter 7, we have developed the quantum mechanical atom-field Hamiltonian in terms of Pauli spin operators and the creation and annihilation operators. We describe the absorption and emission process and define the Dressed State. We consider the rate equation treatment for calculation of absorption coefficients. The Wigner-Weisskopftheory of spontaneous emission is described in detail.

In Chapter 8 we describe theory and experiments of non-linear spectroscopy. The perturbation solution of density matrix equations for a three level atom is used to explain Doppler-free two-photon spectroscopy. We have explained Dark State, Coherent Population Trapping (CPT) and Electromagnetically Induced Transparency (EIT). We also describe experimental results of velocity selective optical pumping and EIT for multilevel rubidium atoms.

Chapter 9 shows how laser radiations can be used to manipulate the velocity of atoms and slow them down. We show that non-cryogenic Doppler cooling method based on lasers can be used to effectively decelerate atoms and control their dynamics. We have given explicit calculation of the radiation force acting on an atom. We also present the operation of a Magneto-Optic Trap (MOT). The experimental techniques of operation of MOT and observation of confined and cooled atom cloud are described at a simple and elementary level. In this chapter we have presented the concept of Bose Einstein Condensation (BEC) and explained how it differs from the usual process of condensation. We attempt to explain how dilute gas of atoms at a very low temperature can undergo transition to a new kind of state, when the atoms lose their quantum identity. It is shown how the process of evaporative cooling in a magnetic trap under the effect of radio-frequency radiation can be used to obtain the temperature in the nano-kelvin region that is necessary to achieve BEC. The observation of BEC and application of BEC are also described.

In the last Chapter (Chapter 10) we provide an introductory idea of quantum computation. We begin with the idea of Einstein, Podolsky, Rosen (EPR) paradox and Bell's inequality. We have introduced quantum entanglement. We also present an idea of laser based quantum computers.

The subject matter of the book is exemplified by a large number of figures including experimental observations and theoretical simulations. This would be useful for clear exposition of observed results and theoretical interpretations. As a book based on fundamental concepts of physics it emphasizes theoretical developments in the subject. We have added a large number of problems in different chapters so that the students can learn the basics and also get an idea of numerical results that can emerge from observations. There are also a number of worked out exercises in the relevant cases.Finally, this book is also helpful to researchers on Laser Spectroscopy or Quantum Optics at the early stage.

I thank Mr Nilanjan Aich for preparing the diagrams presented in this book and also for going through the manuscript. I thank my former students Dr B Ray, Dr S Mandal, Dr D Biswas , Dr. A Pathak and several others by their help. This book would not have been possible without the support and cooperation of my wife Nandita and my daughters Sudeshna and Debalina. I also thank Dr P K Sahu and Dr D Zinder for their active support. Finally, I thank MrMilinda De for the great care taken to publish

Kolkata, 2016 ***Pradip Narayan Ghosh***

Acknowledgement

Catalysed and Supported by the Science and Engineering Research Board

Department of Science and Technology under its Utilisation of Scientific Expertise of Retired Scientists Scheme

Content

Chapter 1
Introduction to Laser

Laser is an intense monochromatic source of electromagnetic radiation. It can generate coherent beam of photons having widely varying frequencies and intensities. LASER is an acronym for **L**ight **A**mplification by **S**timulated **E**mission of **R**adiation. A laser can emit radiation over a large range of power. It can be very small having the size of a micrometer; it can also be as large as the size of a football ground. Laser can emit the light beam continuously in time and it can also operate with pulses of extremely short duration.

1.1 Historical background

The idea of MASER (M stands for microwave), predecessor of laser, was first conceived by C H Townes in 1954. Townes was sitting on a bench in Franklin Park in Washington DC in a mood of contemplation. He was toying with the idea that Einstein's theory of stimulated emission proposed in 1917 might lead to a multiplication of photon number. He visualized that if the process could be effectively continued or sustained there might be an enormous amplification of photon number and hence the radiation intensity. He scribbled something on the back of a used envelope and went back to the laboratory. In a stimulated process as proposed by Einstein an incident photon, incident on an atom with energy level difference that resonates with the frequency ν of the radiation beam, induces or stimulates an atom in the upper energy level to come down to the lower energy level. In this process the atom gives up its additional energy, $\hbar\omega = E_2 - E_1$, by emitting a photon that will have exactly the same energy as that of the incident radiation $\hbar\nu$. Additionally, the phase of the emitted photon will be the same as that of the incident photon thus leading to an efficient multiplication of the energy. The process needs the atoms to be in the upper or excited state and it needs an arrangement so that the process is sustained. These were the basic challenges to Townes and with the help of his students he could design a device that produced amplification of microwave radiation in ammonia molecule. This was the invention of MASER. The idea of light amplification was also conceived by the Russian scientists Basov and Prochorov almost at the same time. Subsequently Townes and Schawlow showed that the same principle could be extended to electromagnetic radiation in the optical region and hence LASER became a plausible source of radiation. The first laser source was developed in 1960.

1.2 Spontaneous and Stimulated Emission and Absorption

An atom has a set of quantized energy levels. It interacts with electromagnetic radiation in a process of energy exchange. The lowest energy state of the atom is the ground state. Thermal distribution of atoms in different energy levels follows the Boltzmann thermal distribution law

$$N = N_0 e^{-E/k_B T} \tag{1.2.1}$$

It describes the number N of atoms with energy E at a temperature T, where N_0 is the number of atoms in the ground state and k_B is the Boltzmann constant. Hence the atoms prefer to occupy the lower energy states. An atom in an excited state will have a tendency to move downwards to an available lower energy level and the extra energy it possessed will be emitted as a photon (Fig. 1.1(a)). This process is known as **Spontaneous Emission**. This photon is emitted in random directions and in random phases. A particle in an excited state spends an average time τ before it emits a photon and steps down to the lower energy level; this time is called the lifetime in the excited state. The life time of a particle in a particular excited state depends on many factors including the particular atomic or bound system, the energy of the state etc.

In case an atom in an excited state with energy E_2 meets a photon with energy $\hbar\omega = E_2 - E_1$, where E_1 is the energy of the lower state, *within the lifetime of the excited state* and if the transition from E_2 to E_1 is allowed by quantum mechanical selection rules then the incident photon will stimulate another photon to be emitted and the atom steps downwards (Fig. 1.1(b)). This process is known as **Stimulated Emission**. The emitted photon will have exactly the same frequency, direction and phase so that the process is coherent (Fig. 1.1(b)).

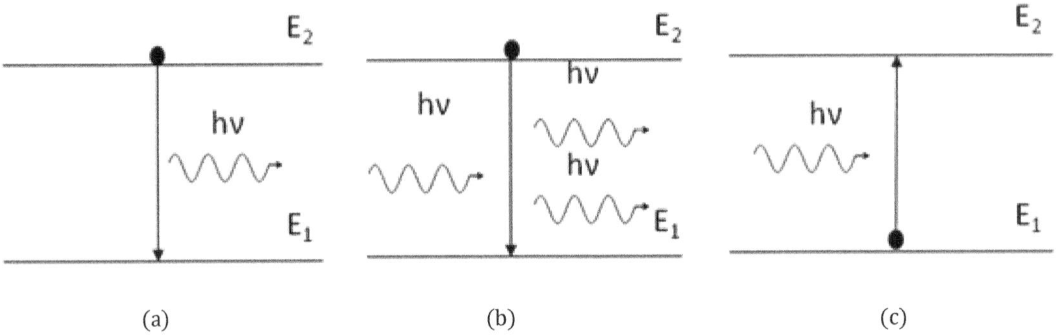

Fig. 1.1 (a) Stimulated Emission, (b) Spontaneous Emission and (c) Absorption, • stands for an atom.

If the atom is in the lower state and meets a photon under similar condition the photon is absorbed and the atom moves to the excited state (Fig.1.1(c)) having higher energy. This is the process of **stimulated absorption**. In fact, absorption is always stimulated. These three basic processes (Fig 1.1) were described by Einstein and will be discussed in Sec. 2.4.

A few points may be noted here:

- The process of stimulated emission is facilitated *if the upper state has a longer lifetime*, otherwise spontaneous emission will dominate over it.
- The rate of stimulated emission is the same as that of absorption. Hence *at thermal equilibrium absorption will always dominate over stimulated emission* since the lower state has a larger number of atoms.
- At room temperature (T ~ 300K) the value of k_BT is approximately 200 cm^{-1} or nearly 6000 GHz. Since the energy difference $E_2 - E_1$ is of the order of 100000 GHz in the visible or optical region, because of Boltzmann law the number of atoms in the upper state is negligibly small compared to that in the ground state. Only in the microwave or lower frequency region the number of atoms in the upper and lower states could be comparable, though the upper state would always have less atoms.

Radiation with frequency ω is incident on the two-level atom. We have assumed a resonance condition so that the radiation frequency is the same as the atomic transition frequency. If N_1 and N_2 be the number of atoms per unit volume in the levels 1 and 2 respectively, the number of atoms absorbed per unit time per unit volume in the transition from 1 to 2 is

$$N_{12} = B_{12} U(\omega)\, N_1 \tag{1.2.2}$$

$U(\omega)$ is the energy density per unit frequency interval. B_{12} is a constant of proportionality indicating the rate of transition.

For the reverse transition from level 2 to level 1 there are two processes. For stimulated emission the rate of downard transition is B_{21}. Hence the number of stimulated emissions per unit time per unit volume from level 2 to level1 is

$$N_{21} = B_{21}U(\omega)\, N_2. \tag{1.2.3}$$

Atoms in the upper state will also undergo spontaneous emission and the this depends on the number of atoms in the upper state and no radiation is necessary. The total number atoms undergoing spontaneous emissions is

$$\Gamma_{21} = A_{21}\, N_2. \tag{1.2.4}$$

For thermal equilibrium of atom radiation field system we need the total number of upward and downward transitions to be the same.

$$B_{12}U(\omega)N_1 = A_{21}\, N_2 + B_{21}U(\omega)N_2. \tag{1.2.5}$$

Hence,

$$U(\omega) = \frac{A_{21}}{B_{12}\frac{N_1}{N_2} - B_{21}}. \tag{1.2.6}$$

According to Boltzmann's law the ratio of populations of the two levels at a temperature T is

$$\frac{N_1}{N_2} = e^{(E_2 - E_1)/k_B T} = e^{\hbar\omega/k_B T}.$$

Hence the radiation density is

$$U(\omega) = \frac{A_{21}}{B_{12}e^{\hbar\omega/k_B T} - B_{21}}. \tag{1.2.7}$$

But Planck's radiation law leads to

$$U(\omega) = \frac{\hbar\omega^3 n_{RI}^3}{\pi^2 c^3}\frac{1}{e^{\hbar\omega/k_B T} - 1}. \tag{1.2.8}$$

Where n_{RI} is the refractive index of the medium and we shall assume $n_{RI} = 1$. A comparison of the two equations show that

$$B_{12} = B_{21}, \tag{1.2.9}$$

$$\frac{A_{21}}{B_{21}} = \frac{\hbar\omega^3}{\pi^2 c^3}. \tag{1.2.10}$$

Thus the rates of stimulated emission and absorption are same. The ratio of the rates of spontaneous and stimulated emissions depends on the frequency of the radiation. The constants A and B are known as the Einstein's A and B Coefficients. At a fixed temperature T, the ratio of the number of spontaneous to stimulated transitions is

$$\frac{A_{21}}{B_{21}\, U(\omega)} = e^{\hbar\omega/k_B T} - 1. \tag{1.2.11}$$

As discussed above, except the radio-frequency and microwave frequency region the radiation frequency is such that $\hbar\omega \gg k_B T$. Hence in all such cases the spontaneous emission dominates over stimulated emission. Since the spontaneous emission is random in phase the radiations from all light sources are usually incoherent.

In absence of any incident radiation the rate of change of the number of atoms in the upper state caused by spontaneous emission is

$$\frac{dN_2}{dt} = -A_{21} N_2. \quad (1.2.12)$$

Hence the time dependence of N_2 is given by

$$N_2(t) = N_2(0)e^{-A_{21}t}. \quad (1.2.13)$$

So the spontaneous emission life time is

$$\tau = 1/A_{21}. \quad (1.2.14)$$

Thus the dimensions of the Einstein coefficients are Hz for A_{21} and $m^3/Jsec^2$ for B_{21} and B_{12}. As per definition of $U(\omega)$ it has the unit of $\frac{J}{m^3}/Hz$. $U(\omega)$ is directly related to the intensity of the incident radiation. The spontaneous emission rate is much higher for the optical region compared to the microwave region.

Exercise 1.1 In the optical region , for the 2P-1S transition of atomic hydrogen the upper state life time is nearly 10^{-9} *sec*.. In the microwave frequency region, for the rotational transitions of a molecule the upper state life time can be of the order 10^{-4} *sec*. Find the Einstein A-coefficient in both cases.

Solution

In the case atomic hydrogen the life time, $\tau = 10^{-9}$ *sec*. Hence Einstein A-coefficient is

$$A_{21} = \frac{1}{\tau} = 10^9 Hz$$

For the microwave transition, the life time, $\tau = 10^{-4}$ *sec*. Hence the Einstein A-coefficient is

$$A_{21} = \frac{1}{\tau} = 10^4 Hz$$

Exercise 1.2 If spontaneous emission life time is 1 nano-second and the transition frequency is 10^7 GHz, find the Einstein's B coefficients.

The spontaneous emission lifetime is $10^{-9} sec$, hence the Einstein A-coefficient is $A = 10^9\, sec$ and the transition frequency is $10^{16} Hz$. The B- coefficient is

$$B = \frac{\pi^2 c^3}{\hbar\omega^3} A$$

Hence,

$$B = \frac{9.85 \times 27 \times 10^{24} \times 10^9}{1.05 \times 10^{-34} \times 10^{48}} = 253.79 \times 10^{19}\, m^3/Jsec^2$$

1.3 Stimulated emission as a process of amplification

In the process of stimulated emission one photon interacting with an atom in an excited state may cause emission of another photon with the same energy and in the same phase. Thus one photon becomes two photons. These two photons may further interact with two other atoms in a similar condition and each of them may cause emission of two further photons (Fig. 1.2). Hence two photons result in four photons in an active atomic medium that have atoms under favorable conditions. This process happens very fast and in a short period of time it may lead to amplification of the photon number and hence enhancement of the light intensity. However, the entire

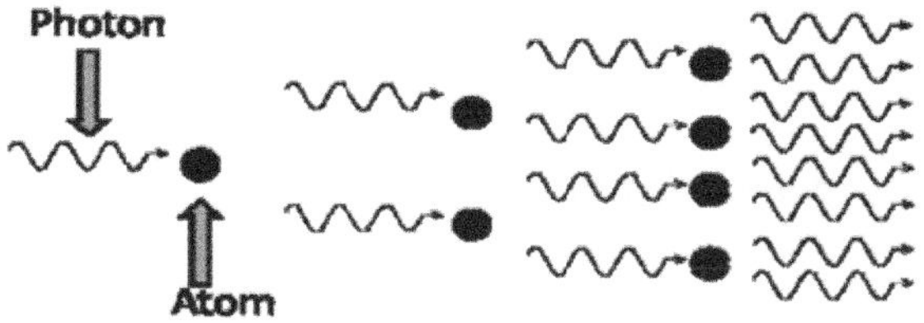

Fig 1.2 One photon incident on one atom produces two photons, this process continues and the photon number multiplies.

process is not as easy as it may appear. For this purpose we need a sufficiently large number of atoms in the excited state and the excited state should be **metastable**, i.e. it should have a long lifetime so that stimulated emission dominates over spontaneous emission. But the thermal distribution law prevents accumulation of a larger number of atoms in the excited state. This is the reason that laser was not possible for a long time even though stimulated emission was known from Einstein's theory. For this purpose one has to create a situation in which we have an atomic medium that has a larger number of atoms in the excited state compared to that in the lower state. This is called **population inversion**.

1.4 Population Inversion for Amplification

Population inversion is achieved in an atomic system if the number of atoms N_2 in the excited state having energy E_2 is greater than the number N_1 in the state with energyE_1, where $E_2 > E_1$. In a thermodynamic equilibrium at a certain temperature T this is not allowed since $T > 0$. If we can create a system where $N_2>N_1$, the Boltzmann distribution law will demand $T<0$. This is a case of thermodynamic non-equilibrium as it is not allowed in equilibrium thermodynamics. The population inversion was possible by pumping a large amount of energy to the system so that a very large number of atoms are transferred to the excited state by a process where a third metastable state can accumulate the atoms for sufficiently long duration so that the stimulated emission can effectively dominate and **amplification** is achieved. This process is described in Fig 1.3 for a three level operation. Atoms are pumped from level 1 to the level 3 by putting in energy to the system. The atoms from level 3 undergo nonradiative transition to the metastable state 2. They are stimulated to emit photons so that the photons emitted in the 2 to 1 transition process multiply to build up the laser radiation. To begin the process of stimulated emission *one photon is necessary to trigger the process.* This could come from any available photon like the photons spontaneously emitted within the enclosure.

Different lasers use different kinds of **pumping** mechanism. In a four level system another level with energy lower than the energy of the level 1 is used. The atoms accumulated in the level 1 may be drained out to this level so that there is no build-up of population in the lower level of the laser transition and the pumping process remains effective.

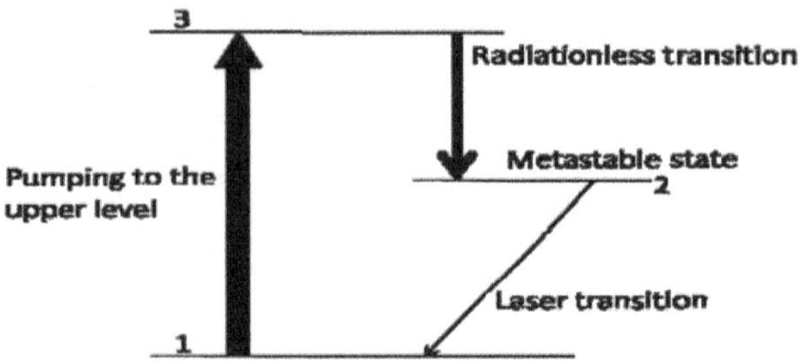

Fig. 1.3 Population inversion in a three level laser.

1.5 Feedback for sustained amplification

As shown in Fig. 1.3, in this process the photon number increases, but the process will stop when the light beam comes to the right hand end of the enclosure containing the atoms (Fig. 1.4) and that can happen in a very short time. In order to sustain the process a part of the energy gained in the stimulated emission process should be fed back to the system so that the multiplication process can continue. This is achieved by keeping the atomic medium confined between two reflectors at the two ends (Fig. 1.4). One of them should be partially reflecting and the other mirror should be fully reflecting. Hence the photons will get reflected at the two ends and will move in opposite directions within the container. If the right hand mirror is 80% reflecting then out of 100 photons incident on it, 20 will pass out of the mirror, but 80 will be reflected back. From the left hand mirror all the 100 photons will be reflected back. This will develop an oscillation within the system. The light energy will increase almost instantaneously. It may appear that such an enormous increase of energy may lead to an explosive situation. But this cannot happen because there are losses in the process.

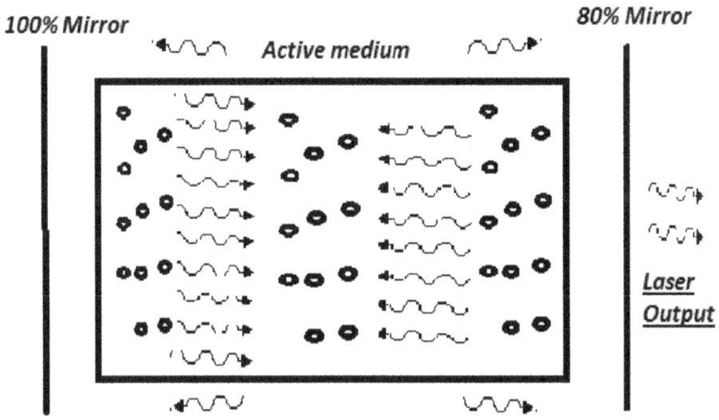

Fig. 1.4 Cavity resonator. The arrows showing photons outside the cavity are losses due to spontaneous emission or imperfect reflection.

The photons that leak out of the right hand mirror constitute the output of the laser. There are also losses if the photon beams are not exactly perpendicular to the mirror and also there are spontaneous emissions.

1.6 Laser components

Laser action needs three essential components: (1) **Cavity resonator,** (2) **active medium** and (3) **pumping mechanism.**

The cavity is confined within two mirrors which reflect the optical wave back and forth within the cavity to help buildup of energy. This provides optical feedback so that the amplified energy is fed back to help further amplification (Fig. 1.4). Thus it acts as an optical oscillator similar to electrical oscillators. Cavity should have *one partially reflecting mirror so that part of the output of the lasing process is transmitted.*

The active medium should be filled with atoms or atomic systems with discrete quantized energy levels such that the population inversion can be established. Medium can be a pure atomic vapor or mixture of atoms like helium and neon, liquid like organic dye or a solid like a diode or crystal of ruby.

The pumping that helps to create population inversion can be done with the help of radiation, electric current, high voltage electric discharge etc.

1.7 Properties of laser radiation

1.7.1 Temporal Coherence

An ordinary source of light emits different wave trains of an average duration τ. There is no phase relationship between different wave trains. In the case of interference fringes from a Michelson Interferometer (Fig. 1.5), if the difference between the time taken by the light to traverse the path from the beam splitter (BS) to one mirror (M_1) and that to the other mirror (M_2) is much smaller than this average time duration, the interference pattern will be

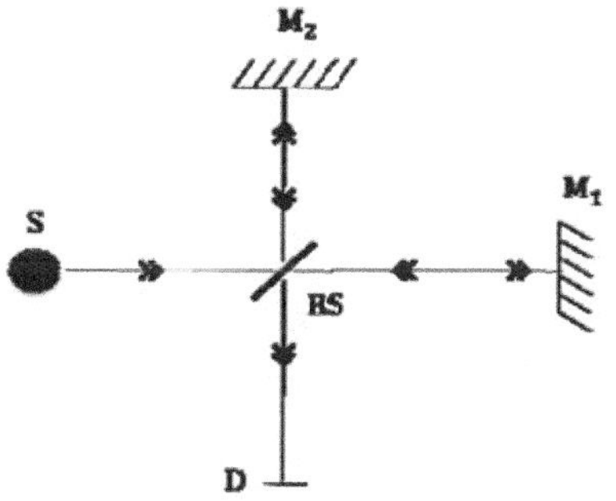

Fig. 1.5 Light beam from the source S is split by the beam splitter BS into radiations that are reflected back by the two mirrors and the interference pattern is recorded by the detector D.

strong. But if this time difference is larger, the interference pattern will be weaker and will gradually disappear. This happens because when the time difference is larger, the two beams of light coming back to the beam splitter after reflection from the two mirrors do not originate from the same wave train (Fig. 1.6) in the source and hence

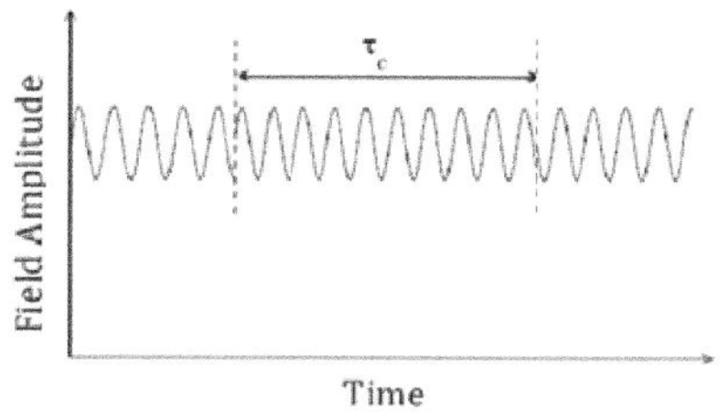

Fig. 1.6 Typical wave train confined between two phase interruptions.

their phase relation will not be known at all. The time duration for which interference pattern is observed is called the **coherence time,** denoted as τ_c (Fig. 1.6) and the length of the wave train $l_c = c\tau_c$ is called the longitudinal **coherence length**. Normally there is no definite distance when the fringe pattern will abruptly disappear. The contrast gradually becomes weak and the fringe pattern eventually disappears. The following exercise shows that the laser can be used for interference experiments with very large path difference.

Exercise 1.3 In the case of Neon light from a discharge lamp the coherence time is nearly 10^{-10} sec, find the maximum path difference between the mirrors for which the fringe contrast will be strong. In contrast for a working laser the coherence time is nearly 2×10^{-3} $sec.$, find the coherence length.

Solution

For ordinary Ne discharge lamp the coherence time, $\tau_c \sim 10^{-10}$ sec, hence the coherence length or maximum path difference between the mirrors for observation of fringes is

$$l_c = c\tau_c = 3.10^{10} \times 10^{-10} cm = 3\ cm$$

For the laser light, the coherence time, $\tau_c \sim 2 \times 10^{-3}$ sec, hence the coherence length is

$$l_c = c\tau_c = 3.10^{10} \times 2 \times 10^{-3} cm = 600\ km$$

A light source does not normally emit at a single frequency, but over a band of frequencies. When the path length is zero or very small the different wavelength components produce fringes superimposed on each other so that the contrast is good. Bur for larger path difference the fringe patterns are displaced from each other so that the fringe becomes weaker. Hence the sharpness of fringes depends on the monochromaticity of the light source. If a light source is a superposition of frequencies within a band of frequencies $\nu_0 - \frac{\Delta\nu}{2} \leq \nu_0 \leq \nu_0 + \frac{\Delta\nu}{2}$, it can be shown that $\Delta\nu = 1/2\pi\tau_c$. $\frac{\Delta\nu}{\nu}$ is a measure of monochromaticity. Thus broader the frequency band smaller is the coherence time.

Exercise 1.4 For ordinary source of light emitting wave length 6000Å, the coherence time is 10^{-10} sec, hence find the width of the frequency band and the monochromaticity.

Solution:

In the given problem, $\tau_c \sim 10^{-10}$ sec, hence

$$\Delta\nu = \frac{1}{2\pi\tau_c} = \frac{1}{2 \times 3.15 \times 10^{-10}} Hz = 1.59\ GHz$$

Since $\lambda = 6000$Å, $\nu = 500{,}000\ GHz$, so the monochromaticity is

$$\frac{\Delta\nu}{\nu} = \frac{1.59}{5\times10^5} = 3.18 \times 10^{-6}.$$

For the laser beam this factor $\frac{\Delta\nu}{\nu}$ will be much smaller, since $\Delta\nu$ can be of the order of a few kHz. Thus the known monochromatic sources of light like the sodium vapor lamp do not emit strictly monochromatic radiation. In the case of ordinary sources of light, the emitted radiation is spontaneous emission of light, which has a random phase.

Exercise 1.5 Find the maximum bandwidth for a laser source with coherence time 10^{-3} sec. Find a measure of monochromaticity for the laser wavelength 6328 Å.

Since $\tau_c \sim 10^{-3}$ sec, the bandwidth is

$$\Delta\nu = \frac{1}{2\pi\tau_c} = 159.23\ Hz$$

Laser wavelength is $\lambda = 6328$Å, so $\nu = 4.74\ \times 10^{14}\_Hz$

Hence the monochromaticity is

$$\frac{\Delta\nu}{\nu} = 3.37\ \times 10^{-13}$$

Theoretical treatment of coherence properties of electromagnetic radiation will be presented in Chapter 6.

1.7.2 Spatial Coherence

If we consider any two points S_1 and S_2 that lie on the sane wave-front of an electromagnetic wave at time $t = 0$ and if the electric field of the electromagnetic wave at these two points be $\mathcal{E}_1(t)$ and$\mathcal{E}_2(t)$, the phase difference of the wave at these two points will be zero. If the phase difference remains zero at some later time t, we can say that the two points have perfect spatial coherence. If this happens for any two points of the wave-front the wave has perfect spatial coherence. The point S_2 must lie within a finite area (say, S_c) around the point S_1 in order to have a good phase correlation. Within such an area we can say that the wave has partial spatial coherence.

In the case of a Young double slit experiment the light beams originating from a point on the source which is equally distant from the two slits will have zero path difference and the phase difference is zero so that perfect fringe system may be formed. This happens for the axial point. But the light beams from any point S' away from this axial point will have a non-zero path difference. If the path difference is equal to λ/2 the source produces an interference minimum at the detector. It can be shown that good interference fringes can be observed if

$$l \ll \lambda D/2, \qquad (1.7.1)$$

where l is the perpendicular distance SS' of the point from the axis passing through the centre of the double slit, D is the distance of the double slit (Fig. 1.7) from the light source and d is the distance between the slits (Problem 1.4).

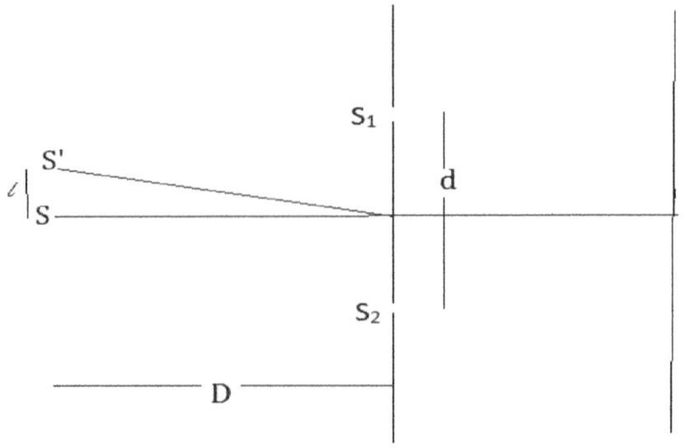

Fig. 1.7 Young's double slit experiment with a source S' a distance l above the axial point S. The interference pattern is formed on the screen on the right. If $l \sim \lambda D/d$, the interference fringes will disappear.

In the case of ordinary light source we should pass the light through a pin-hole in order to see the fringes, since the ordinary light source is incoherent. In contrast a laser beam is highly spatially coherent.

We should note here that there is a fundamental difference between temporal and spatial coherence and they are independent of each other. A light bean with high spatial coherence may have poor temporal coherence.

Exercise 1.6 In a Young's double slit experiment (Fig. 1.7) find the maximum distance of a point from the axis for which good interference fringes can be observed for sodium light source placed at a distance of 20 cm from the double slit and the distance between the slits is 0.01mm.

Solution:

Distance between the two slits $d = 0.01\ mm = 10^{-5} m.$

Wavelength of sodium light $\lambda = 589.3\ nm$

Distance between the source and the double slit D = 20 cm=0.2 m.

Hence the maximum distance of a point from the axis for observation of good fringes

$l = \lambda D/d = \frac{589.3\times10^{-9}\times0.2}{10^{-5}}\ m = 1.179\ cm.$

1.7.3 Directionality

Directionality of the laser radiation results from the fact that the laser beam originates from the multiple oscillations of the radiation within a cavity. When this beam comes out of the mirror that has partial transparency it is expected to be perpendicular to the mirror of the output radiation. Hence it should go out as a pencil of radiation. However, the opening of the mirror is finite in size. So the radiation beam making small angle with the direction perpendicular to the mirror may also consist of a part of the output beam. This can be shown in the figure 1.8. A monochromatic beam with uniform intensity and having a plane wave-front falls on the aperture having width *d*. As per Huygen's principle the wave front at a point P on a screen at a certain distance from the aperture A will be superposition of all the wave-fronts emitted from each point of the aperture. Because of the finite size of the aperture the emitted wave will be diffracted. In the case of perfect spatial coherence the beam has finite angle of departure θ from the normal direction. The angular spreading of the light beam having wavelength λ is given by the diffraction condition

$$\theta = \lambda/d. \tag{1.7.2}$$

In the case of partial spatial coherence, the diffraction is larger and the beam is less directional. In this case d is replaced by ${S_c}^{1/2}$ (a finite area as mentioned above) in the equation (1.7.2). The laser beam is usually so focussed that it is diffraction limited so that we can get a beam with least divergence and have perfect spatial coherence. Thus high directionality is a result of high spatial coherence of the laser radiation.

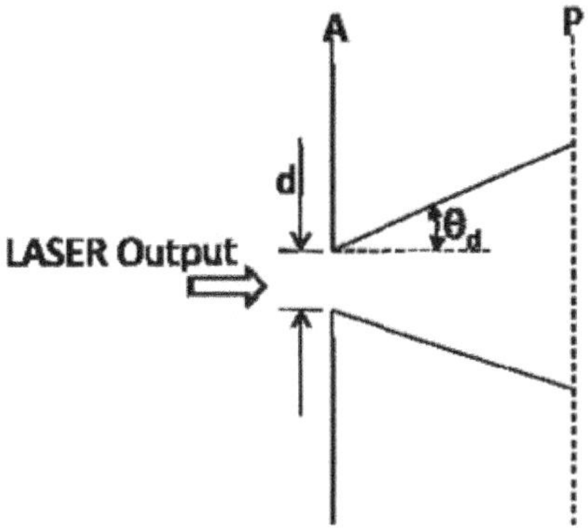

Fig. 1.8 Diffraction of laser beam through an aperture of width d.

Exercise 1.7 For perfect spatial coherence find the angular aperture from the normal direction for an aperture of width 0.1 mm and light wavelength 6000 Å.

Solution:

Angular aperture

$$\theta = \lambda/d \quad = \frac{6\times10^{-7}}{10^{-4}} = 0.006\, rad$$

1.8 Types of lasers

In this section we first point out the difference between cw and pulse operation of lasers. Then we shall discuss the principle of operation of some of the cw lasers like He-Ne laser which is a gas laser and the solid state semiconductor diode laser the liquid state dye laser. Depending on the active medium the operation principles of lasers are very much different from each other. The principles of working of Titanium Sapphire and gas lasers like Carbon Dioxide, Argon Ion etc. have been described in detail in literature. We shall not discuss them. The readers may consult the book by Yariv (Ref. 4) or by Quimby (Ref. 2).

1.8.1 cw and Pulse Lasers

As per discussions so far once a laser source is switched on and the laser acts above threshold it will emit radiation and the emission is continuous as long as the laser remains switched on. Our discussion in this book is mostly confined to this kind of lasers called continuous wave or cw lasers. But using special techniques we can also produce laser pulses. In such cases either pump rate or cavity loss or both can be time dependent. The transient behaviour may also be obtained by inserting a non-linear optical element inside the cavity such that this non-linearity can produce a departure from the cw operation. It is possible to produce a pulse whose duration is of the order of the inverse of the line width of the emitting transition. In the case of gas lasers the line width is normally very small so the lasers pulses can have duration of the order of nanosecond. Such nanosecond pulses can also be obtained from flash lamps. But in the case of solid state lasers the line width is much larger and hence pulse duration can be extremely small like femto-second to atto-second. For single mode laser relaxation oscillation, Q-switching or gain switching may be considered. In the case of multimode lasers a special technique called mode locking is used. The operation of the pulse laser is discussed in sec. 1.11.

Short pulse duration is analogous to the monochromatic property of the radiation. For a monochromatic wave the energy is concentrated in the frequency space, whereas short pulse lasers are concentrated in the time space. But there is a fundamental difference. All lasers can be monochromatic. But all lasers cannot have very short pulse duration. Solid and liquid state lasers can produce very narrow pulses as they have broad bandwidth.

1.8.2 Gas Laser: Helium Neon Laser

The gas laser popularly called He-Ne laser was the first gas laser operated successfully by Ali Javan in 1961. It is one of the simplest and cheapest lasers producing low power visible light. He-Ne lasers can actually emit four laser radiations at wave lengths 6328Å (red), 5430 Å (green), 33910 Å (IR) and 11520 Å (IR). The green laser radiates at a lower power than the red light. The first gas laser action was achieved at the wavelength 11520 Å.

Gas atoms have discrete quantized energy levels. The levels are very narrow and sharp. Hence the emitted radiation can be highly monochromatic. Electrical discharge of gas atoms is used to have effective population inversion. In the He-Ne laser a gas mixture of 1 Torr Helium and 0.1 Torr Neon (1 Torr= 1 mm of Hg) is the active medium enclosed in a cylindrical cavity of length nearly 80 cm and diameter 1 cm (Fig. 1.9). The ends of the discharge tube are fitted with Brewster windows. The tube is confined by two mirrors that form the reflectors in the cavity, one of them is fully reflecting and the other is partially reflecting and provides the output beam. The Ne atoms are the actual lasing atoms, He atoms are used for pumping to the upper levels of Ne by collision process.

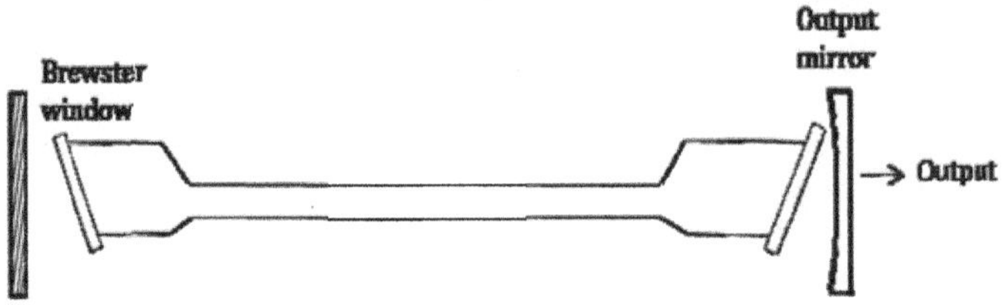

Fig. 1.9 Typical He-Ne laser tube.

The energy levels of He and Ne atoms involved in the laser action are shown in Fig. 1.10. Helium has two electrons with the ground state as 1^1S, the two electrons have opposite spin angular momenta and both have zero orbital angular momenta. When one of the two electrons is excited to the upper state $n = 2$, it can have either the same spin (S=1) or opposite spin (S=0) as compared to the other electron. This results in two excited states 2^3S and 2^1S. The Ne atom has ten electrons with the configuration $1s^2 2s^2 2p^6$. It also has the ground state1^1S. This is true for all inert gas atoms with completely filled shell. With one electron excited to the upper level we can have a number of excited states as shown in Fig. 1.10. The levels are further split because of the interaction with the remaining 5 electrons in the 2p orbital.

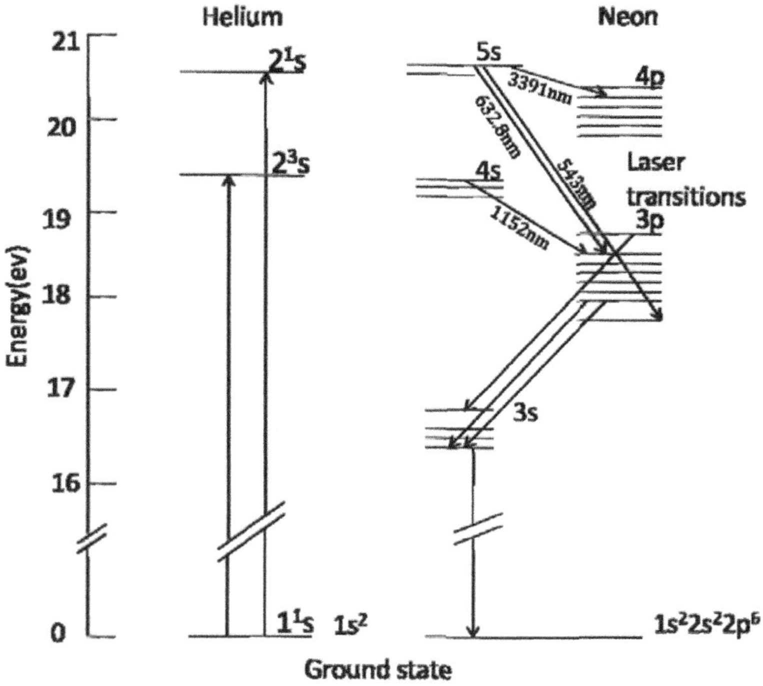

Fig.1.10 Energy levels of He and Ne atoms showing pumping of He atom to 2S levels, resonance transfer to 4s and 5s energy levels of singly excited Ne atom and stimulated emission to show three lasing transitions.

When an electric discharge is produced in the gas mixture, the electrons are accelerated along the axis of the tube. They collide with the Helium atoms and excite them to the higher energy levels. The excited states of He are metastable. The upper 2^3S state has a life time of 10^{-4} sec and the upper 2^1S state has a life time of $5x10^{-6}$ sec. The $S \rightarrow S$ transitions are dipole forbidden and triplet-singlet transitions are spin forbidden. It is found that the excited energy levels of Ne are close in energy to the excited He atoms. So there is resonance energy transfer by collision. When a helium atom in the 2S level collides with ground state Ne atom, energy is transferred to the Ne atom. So they are excited and the He atom is de-excited to the ground level. The long lifetime of excited states of He makes the process very efficient. Since the gas pressure of He is ten times higher than that of Ne, the selective excitation to these Ne atomic levels dominates over excitation of Ne atom to other levels. The electrons thus excited to the 4s or 5s levels of Ne have population inversion with respect to the lower lying levels of 3p or 4p. There are dipole-allowed transitions to the lower levels at wavelengths 1152, 3391, 632.8 and 543 nm from these upper levels. They are suitable as laser transitions. The population inversion can be maintained because the lifetime (~100 ns) of the upper s level is much higher than the decay time of the lower p levels (~10 ns). The electron excitement rate to the p states by electron-Ne collision in the discharge is also small. Specific laser oscillation frequency can be selected by employing mirrors that can reflect over narrow band of frequencies around the centre of the frequency of interest. The electrons from the lower laser levels (p) are de-excited by spontaneous emission to the 3S level and finally to the ground level by collision with wall and also by atomic collisions. The laser emission at 632.8 nm is Doppler broadened with a width of nearly 1.5 GHz. The natural broadening $\Delta\nu_N = \frac{1}{2\pi T_n} \cong 17MHz$ is determined by the life time of the s and the p states. Collision broadening at the operating gas pressure is also small ($\sim 1\ MHz$). The Doppler broadening of the Infrared lines are smaller because of the lower emission frequencies. Some characteristic features of He-Ne laser operation are given in Table 1.1

Table 1.1 Some typical data of He-Ne laser operation

Upper state Lifetime	Lower state Lifetime	Transition Line width	Power
150ns	10ns	1.5 GHz	1-50mW(cw)

Brewster windows may be used to obtain polarized light beams. Being highly directional, monochromatic and having low power and operating on a single mode He-Ne lasers have found wide applications in alignment, bar code scanners and holography as well as for various metrological purposes. The 632.8 nm red light is more popular for applications. The green light at 543 nm has lower power, but because of better visibility to eye it is being used for alignment and cell cytometry. Since the diode lasers are much smaller in size and much more convenient, the gas laser are being replaced by laser diodes.

1.8.3 Semiconductor Laser

Semiconductor lasers constitute one of the most important types of lasers that have found wide ranging applications. In this laser the active medium is a direct band gap material. We cannot use semiconductor elements like Si or Ge. Instead we have to use a combination of Group III elements (like Al, Ga, In) and Group V elements (like N, P, As). Examples are GaAs or ternary compounds like AlGaAs, InGaAs or quaternary compounds like InGaAsP. They emit over the 600 to 1600 nm region. More recently diode lasers like InGaN have been found to emit in the visible blue region of wave length. The wide band gap semiconductors consisting of combination of Group II (such as Cd and Zn) and Group VI (such as Si, Ge) elements emit in the blue to green region of the spectrum. The small band gap semiconductors of Group IV-VI compounds like Pb salts of S or Se emit in the mid infrared region. But they need cryogenic cooling because of the small band gap. These lasers are now replaced by quantum cascade lasers that operate at room temperature for the same wavelength region.

In a Light Emitting Diode (LED) light is emitted by radiative recombination of electrons and holes by a process of spontaneous emission. In a semiconductor laser light is generated by a process of stimulated emission In this process the probability of emission depends on the number of photons already present or the light intensity. Additional photons are created by those already present, so there is amplification. This is sustained by adding reflective elements. At low input power the diode works as an LED, the laser emission starts above a threshold power or input current to the diode. The laser power W can be expressed as

$$W = \alpha\,(i - i_{th}). \tag{1.8.1}$$

Where i_{th} is the threshold value of input current above which laser emission starts. This is depicted in Fig. 1.11. Below threshold current the diode emits a low power by spontaneous emission, so it exhibits LED operation. Above threshold current it behaves as a laser and the power increases linearly. The emitted photons are all emitted into the same mode and propagate in the same direction. Typical value of the slope is $\alpha = 1$ mW/mA, hence if $i_{th} = 20$ mA and $i = 80$ mA the output power is 60 mW.

In laser diodes double heterostructure junctions are normally used. There is one junction between n-type (AlGaAs) and a lightly doped p-type (GaAs) and another junction between this lightly doped p-type and a more

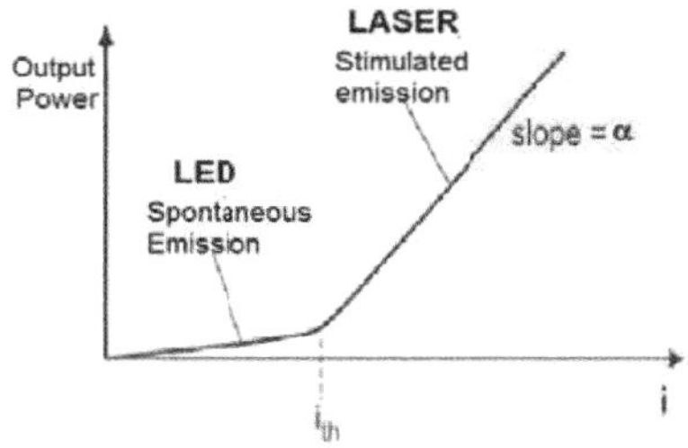

Fig. 1.11 Output power of a semiconductor laser showing threshold current at which laser emission starts.

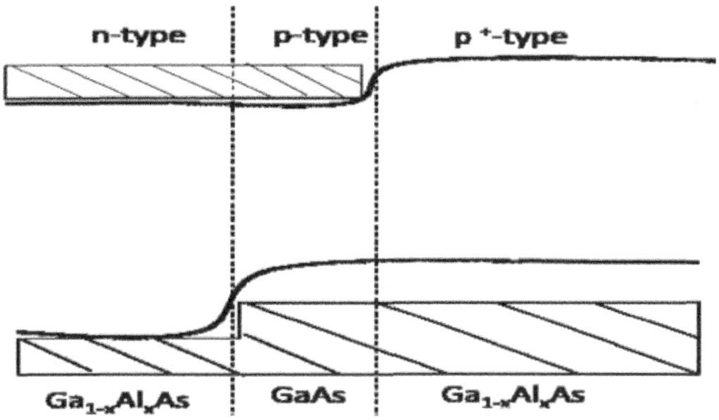

Fig. 1.12 Band structure and emission by recombination in a double heterostructure diode laser.

highly doped p-type (AlGaAs) semiconductors. The energy bands (Fig. 1.12) in the different semiconductors shift due to the band gap differences in the materials and also the p-n junction potential. The gain region is confined to the middle region with a proper choice of the material and doping. These lasers have high optical confinement. Electrons and holes enter the active region of thickness d (Fig. 1.13). With forward bias applied to the p-n junction the electrons in the n-type region are injected into the sandwiched p-type region. With highly doped p-type semiconductor of wide band gap on the other side of the p-n junction, the electrons are confined to the active region and this helps build up population inversion and high light emission intensity. The active layer of small thickness 0.05 to 0.2 μm is GaAs or $Al_yGa_{1-y}As$ with a small y value. The n- and p-type semiconductors on the two sides of the active layer are Al_xGa_{1-x} As with $x > y$. For x =0.3 the band gap of the active layer is 1.8 eV, whereas the band gap of GaAs is 1.4 eV. The refractive index of GaAs is higher than that of $Al_xGa_{1-x}As$, so the light remains confined in the layer by total internal refraction. If we define the number of excess electrons in the recombination region as N, we can find the rate equation for the electrons in the presence of an injection current i that may also vary in time. There are two processes:- (1) addition of electrons induced by the injection current and (2) the decay of the number of electrons due to recombination of electron-hole pair. Hence the rate equation can be written as

$$\frac{dN}{dt} = \frac{i}{e} - \Gamma N, \tag{1.8.2}$$

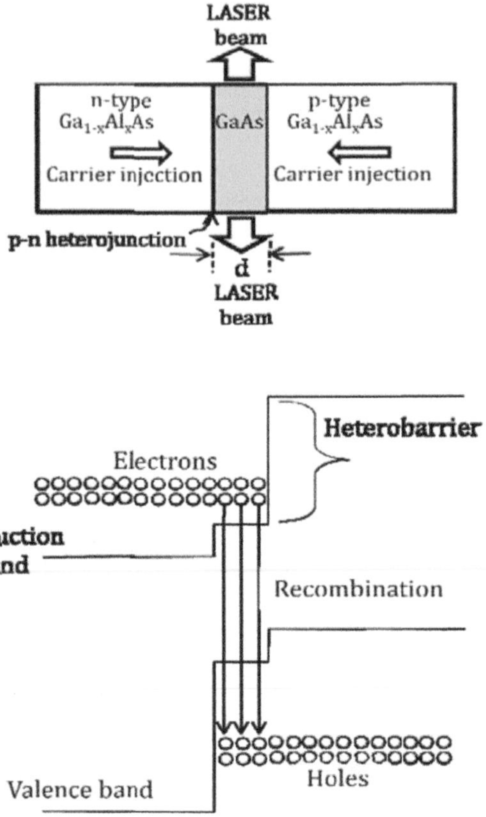

Fig. 1.13 The upper diagram represents the active region in a double heterostructure laser diode. The lower diagram shows the electron-hole recombination process in the active region.

where Γ is the probability of decay of a single electron per unit time in the active region by recombination effect. If $i = 0$, we get,

$$N(t) = Ce^{-\Gamma t} = C\, e^{-t/\tau}, \tag{1.8.3}$$

with C as a constant and $\tau = 1/\Gamma$ as the electron lifetime. Thus when the current is switched off, the active electrons and hence the light emission decays exponentially. In the case of steady state $\frac{dN}{dt} = 0$. So we get at threshold

$$\frac{i}{e} = n\, Ad/\tau, \tag{1.8.4}$$

where n is the electron density per unit volume, A is the cross section of the junction and d is the width of the active layer and hence $N = n\, Ad$. In the case of radiative transition the number of electrons per unit volume is the same as the number of holes per unit volume. In such cases the probability of decay Γ is proportional to the electron or hole density. Hence $\Gamma = Bn$, so the threshold current is

$$i = eAdn^2B, \tag{1.8.5}$$

where B is the probability of recombination per unit time per unit hole density, it has a value of the order of $10^{-11}\ cm^3/sec$ for GaAs. The threshold current density is obtained as

$$I_{th} = \frac{i}{A} = edn_{th}{}^2B, \tag{1.8.6}$$

n_{th} is the density of electrons at threshold required to achieve lasing. It should be noted that for lower threshold current density the width d of the active region should be smaller. However if d is too small, the current profile may extend beyond the active region leading to loss of amplified light intensity. An optimum width of d =0.1 μm is required for the usual heterostructure diode lasers. In order to minimize the actual current one must also use a lower cross sectional area. This needs to be optimised so that sufficient space is available for photon amplification and have good intensity.

The diode of small size of the order of 300 μm is encapsulated in a casing of nearly a few mm in size with projected leads. The laser operation needs a low current in the range 80 to 120 mA. The laser also needs a temperature and a current controller. Usually a Peltier element is used for thermoelectric heating or cooling that can scan the frequency over a broad range. The current control provides fine tuning of the laser frequency. In order to obtain narrow laser line frequency an External Cavity system is used, as discussed in Chapter 4.

Because of low cost, simple operation, high monochromaticity and low intensity the diode laser finds widespread applications in telecommunications and computer hardware. It has no substitute in communication of signals, compact disc readers, laser pointer and alignment work. Because of low power it is not of much use in heavy industry. It is available normally in the near infrared or red region, although blue light emitting diode lasers are also available now-a-days.

Exercise 1.8 A double heterostructure laser diode has an active cross section of $1\ \mu m^2$ and width 0.5 µm. Find the threshold current and threshold current density. Assume threshold electron density to be equal to 10^{18} per c.c.

Solution:

Threshold current is

$$i = eAdn^2B$$

In this problem, $A = 1\ \mu m^2$, $d = 0.5\ \mu m$

$n = 10^{18}$ per c.c, $B = 10^{-11}\ cm^3/sec$

$e = 1.602\ \times 10^{-19}\ Coulomb$

Hence the current

$$i = 1.602 \times 10^{-19} \times 10^{-8} \times 0.5 \times 10^{-4} \times 10^{36} \times 10^{-11}$$

$$= 0.8 \times 10^{-6}\ Coulomb/sec$$

Threshold current density is

$$\frac{i}{A} = 80\ amp/\ cm^2$$

1.8.4 Quantum Well Laser

For double heterostructure lasers we need a minimum width of the active region. If the width is reduced much further to 10 nm region quantum wells are formed (Fig. 1. 14). There are discrete energy levels like particle-in-a-box system. The movements of the electrons or holes perpendicular to the layers is constrained. They move parallel to the layers virtually in two dimensions. The electron-hole recombination probability depends on the

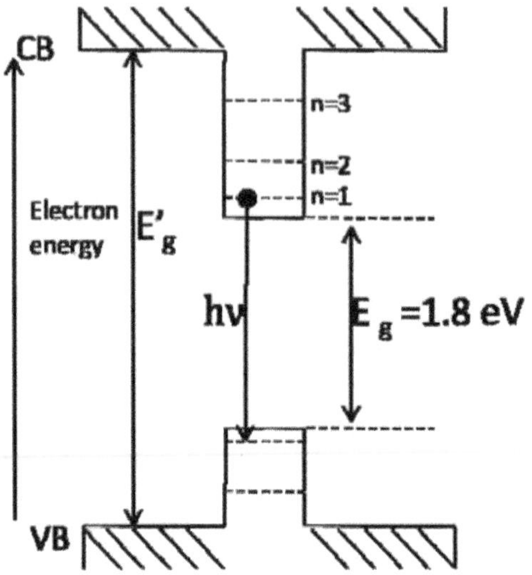

Fig. 1.14 Energy levels in a quantum well structure; they depend not only on the band gap but also on the width d.

number of energy levels available close to the transition energy. In a quantum well laser this number density of states is large as the energy levels are well defined. Hence the gain intensity can be large and this can compensate the loss due to the extremely small size of the active layer. One method to increase the optical confinement is to add more quantum wells in order to produce multiple quantum well (Fig. 1.15) so that each has a confinement of the same optical mode thereby increasing the net gain. The Multiple Quantum Well (MQW) needs more threshold current since each individual well needs injected electrons. This results in higher gain and faster time response. These laser frequencies can be chosen very easily by selecting the value of d and the material. They need smaller threshold current (Eq. 1.8.6) because of having small d. So there is flexibility to increase or multiply the number of layers. They have many advantages over the DH laser diodes.

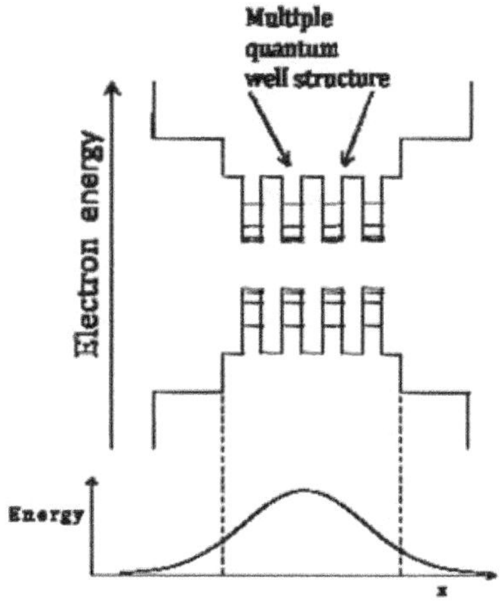

Fig. 1.15 Multiple quantum well and energy profile.

1.8.5 Dye Laser

The liquid state dye laser is the most widely used tunable laser in the visible region. It is based on solution of an organic dye in a liquid solvent like methyl or ethyl alcohol, water or glycerol. The usual organic dyes are long chain polyatomic molecules like Rhodamine 6G (Fig. 1.16(a)) with emission wavelength at 590 nm or Coumarine (Fig. 1.16(b)) with typical emission wavelength at 450 nm. Electronic energy levels of a dye molecule are characterised by a set of singlet (S=0) and triplet (S=1) states. Each of these states has a stack of vibration and rotation sublevels (Fig. 1.17). The ground singlet state of the organic dye is S_0 and the first excited singlet state is S_1, while the excited triplet states are T_1 and T_2. Since the solvent molecules suffer a strong collision with the dye, the transition across the energy states exhibits almost a continuum.

When the dye molecules are subject to electromagnetic radiation there are transitions following the selection rule $\Delta S = 0$. Thus only singlet-singlet and triplet-triplet transitions are allowed and singlet to triplet transitions are forbidden. In the presence of the radiation there will be absorption to raise the molecules from the ground state S_0 to a vibrational level of the S_1 state following the Franck-Condon principle. Since the vibrational and the rotational levels are very closely spaced, they are unresolved and this will show a very broad absorption band. The dye molecules have a large transition dipole moment. This arises from the fact that the π −electrons can move freely along the entire chain of the molecule.

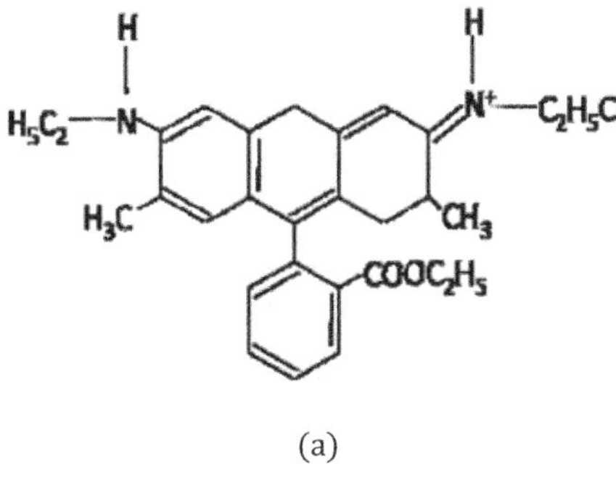

(a)

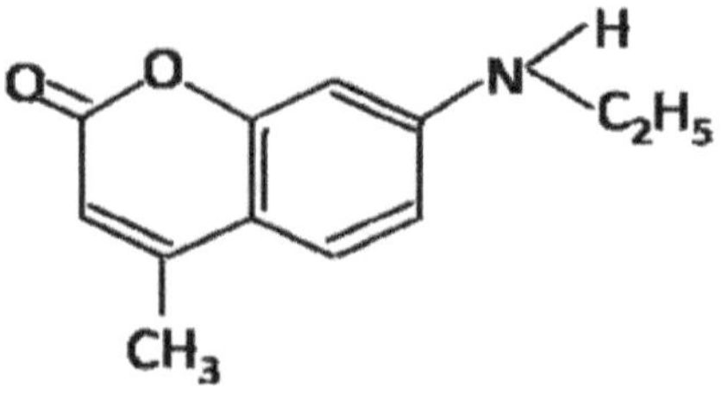

(b)

Fig. 1.16 Chemical structure of common organic dyes , (a) Rhodamine 6G and (b) Coumarine 2.

Hence the absorption cross-section will be very large, of the order of 10^{-15} cm^2. They will go through fast non-radiative decay from the excited state (Fig. 1.17) to the ground vibrational level of the S_1 state in a time of 100 fs because of collisional deactivation. From the ground state of S_1 there will be radiative transition (emission) to some of the upper vibrational levels of the S_0 state again following the Franck-Condon principle.

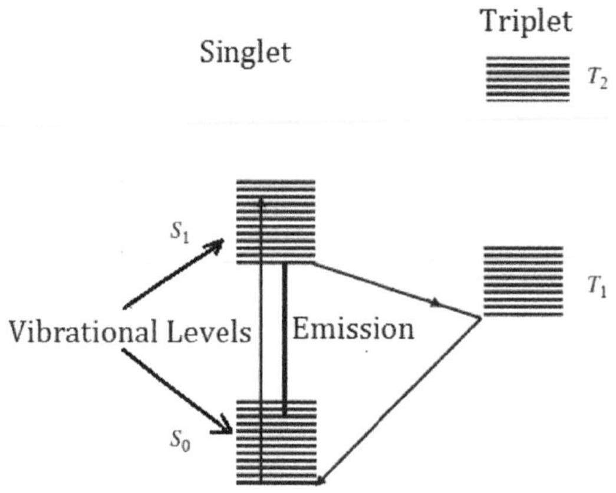

Fig. 1.17 Energy level diagram of Rhodamine showing absorption and stimulated emission along with collisional transfer to the triplet states.

The fluorescent emission is also broad without any feature. Since the dipole moment is strong the emission spectrum is also strong. Once they reach the vibrational state of the lower singlet state, they will have fast non-radiative decay to the ground vibrational state of S_0. When the molecules are in the lower vibrational state of S_1they can have transition also to T_1 by a process called intersystem crossing. This is a radiatively forbidden transition that may occur via collisional process. There may also be transition from T_1 to S_0 by collision induced near-resonant energy transfer by collision with other species in the solution provided the total spin of the interacting partners is preserved. The molecules in the lower levels of T_1 may also absorb radiation to go to the levels of T_2. This absorption occurs in the same wavelength region as the laser radiation. Hence this is a disadvantage for the laser action.

The fast non-radiative decay within the S_1 state is effective to populate the upper laser level and at the same time the fast non-radiative decay within the ground electronic state S_0 to the vibrational lower level is effective to depopulate the lower laser level. Thus the two processes help achieving and maintaining population inversion and the system acts as a four-level laser.

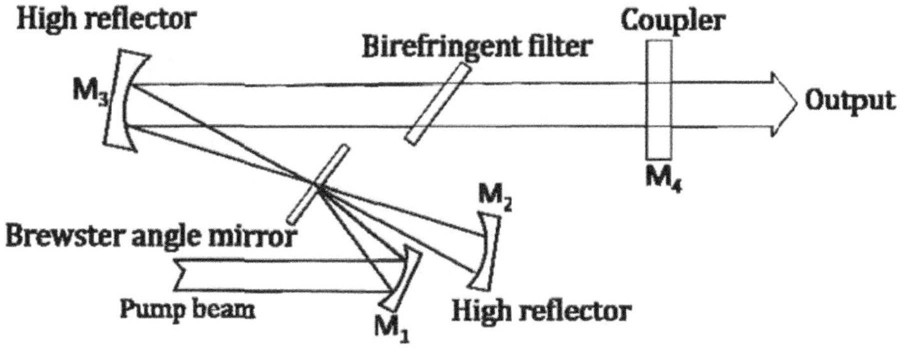

Fig. 1.18 Gas laser pumped dye laser system. The reflector M_3 has higher radius of curvature than the reflector M_2. The pump is usually a gas laser like Argon ion laser.

For continuous pumping of dye laser Ar^+ or Kr^+ laser radiations are used. Laser is tuned by using a birefringent filter (Fig. 1.18) placed inside the laser cavity. In order to achieve single longitudinal mode operation a birefringent filter and Fabry-Perot etalons are used. The laser beam can be linearly polarized with the birefringent filter.

Dye lasers can also be operated in pulsed form by using fast and intense flash lamp or another pulsed laser. The pulse duration should be very small. Pulsed nitrogen lasers have been used for pumping many dyes operating in the visible region. To obtain higher power excimer lasers, copper vapour laser or G-switched Nd:YAG lasers have been used to obtain pulses of short duration from the dye laser.

Organic dye lasers have found many applications in scientific research because of wide spectral coverage, wavelength tunability, high power, and also possibility of generating femtosecond or picosecond pulses. They have also found large application in biomedical field.

1.8.6 Solid State Ruby Laser

Solid state lasers have as their active species ions introduced in a transparent host material in crystalline form. Ruby laser belongs to this category. Semiconductor lasers discussed earlier are not called solid state lasers, because they have a different pumping mechanism and different type of laser action. The first laser that was operated successfully was the Ruby laser developed in 1960 by Maiman. There are many other solid state lasers that are doped crystals in which some of the ions are replaced by ions of the same group.

Ruby is a precious crystal consisting of aluminium oxide in which some of the aluminium atoms are replaced by chromium atoms. The energy levels of chromium are used for laser action. Ruby laser is a three level laser with the chromium energy levels (or bands) shown in Fig. 1. 19.

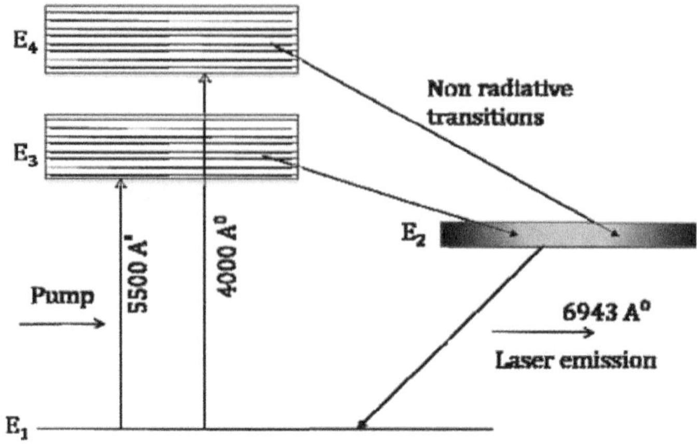

Fig. 1.19 Pumping and laser action in Ruby laser

The pumping or population inversion is achieved by using a flash lamp like Xenon or Krypton lamp emitting in the region 5500 or 4000 Å. As a result of pumping the atoms of chromium are raised from the ground level E_1 to the energy bands E_3 or E_4 . These states have a short life time of 10^{-8} or 10^{-9} sec. So the atoms quickly go through non-radiative transition to the level E_2, which is a metastable state having a long life time of a millisecond. The stimulated emission from level E_2 to the ground level E_1 gives rise to laser emission at wavelength 6943 Å . Since the flash lamp operates in pulses the laser emissions are in the form of pulses. When the flash lamp is on, there is buildup of population in the upper level. The laser action starts, but as soon as the flash goes off, the upper level population is depleted and the lasing action stops. This process continues. The intensity of the laser output varies randomly during the operation giving rise to spikes.

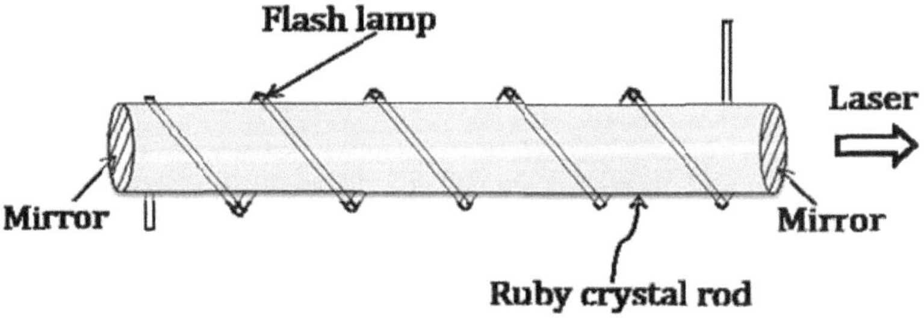

Fig. 1.20 Schematic diagram of a Ruby laser

In a typical Ruby laser the flash lamp is helical in shape and surrounds the ruby rod having reflectors at the end (Fig. 1.20). The Ruby rod is usually 10 cm long and has a diameter of the order of 1 cm. The input energy required for excitation is nearly a 10-20 kJ. Efficient focussing of the light energy may also be achieved by using an elliptic reflector in which the Ruby rod and the flash lamp are placed at the two foci of the ellipse.

Ruby laser has the advantage of a long life time and narrow line width. They operate in the visible region. Hence they have a range of practical applications. They are used in holography, laser ranging etc. It is one of the most important practical lasers.

1.9 Laser Beam Propagation

1.9.1 Modes in Optical Resonator

We discussed in section 1.5 that laser beam is confined in an optical resonator for effective feedback of the radiation. The simplest arrangement of an optical resonator is a pair of mirrors reflecting the light beams. The resonator has an important effect in shaping the frequency spectrum. If two mirrors are separated by a distance L, the light rays that are moving parallel to the axis or perpendicular to the mirrors can participate in the feedback process. Other beams are lost. The forward and reflected light beams will reinforce each other if the phase of the beam after propagation through a distance 2L remains unchanged, or the field satisfies the boundary condition $\mathcal{E}(x+2L) = \mathcal{E}(x)$. The distance $2L$ should be an integral multiple of the wavelength. This leads to the wavelength condition

$$n\lambda = 2L, \quad (1.9.1)$$

where n is an integer. The frequency of the n-th mode in the cavity is then given by

$$\nu_n = \frac{nc}{2L}, \quad (1.9.2)$$

where the refractive index of the medium is considered as unity. These frequencies are known as the mode frequencies of the cavity. There is resonance enhancement or the electric field has a large amplitude at these frequencies.

Physically it means that the optical energy can be stored only at these frequencies. The frequencies are equally spaced and form a comb spectrum (Fig. 1.21) with a spacing of $\frac{c}{2L}$. If two oppositely running waves have a frequency ω and a constant phase ϕ that satisfy the boundary condition at the walls, the electric field is given by

$$\mathcal{E}(z,t) = \mathcal{E}_0 \cos(kz + \omega t + \phi) + \mathcal{E}_0 \cos(kz - \omega t + \phi) = 2\mathcal{E}_0 \cos(kz + \phi) \cos \omega t. \quad (1.9.3)$$

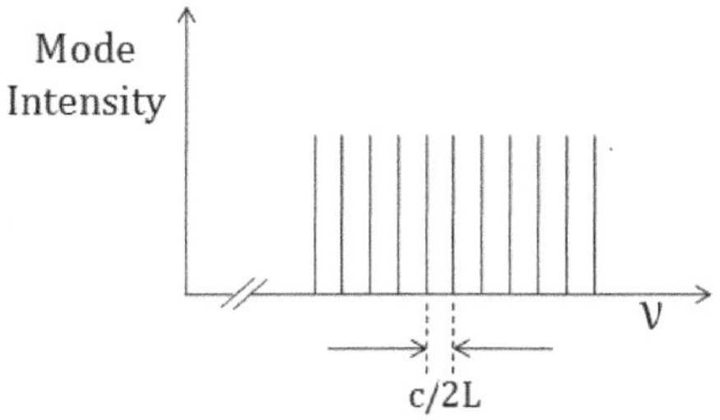

Fig. 1.21 The Comb spectrum with each peak having a sharp intensity and zero width.

This is the standing wave with nodes at the two mirrors (Fig. 1.22).

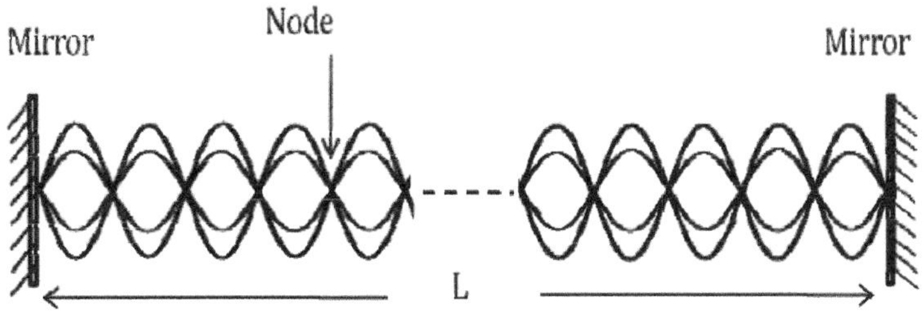

Fig. 1.22 Standing wave enclosed in a cavity.

Exercise 1.9 Find the mode spacing and the number of modes in He-Ne Laser oscillating at 632.8 nm in a cavity of length 30 cm.

Solution:

Wavelength $\lambda = 632.8\ nm$, length of the cavity, $L = 30\ cm = 0.3\ m$.

The number of modes,

$n = 2L/\lambda = \frac{2\times 0.3}{632.8\times 10^{-9}} = 948167$

Mode spacing is

$\frac{c}{2L} = \frac{3\times 10^{8}}{0.6}$ Hz= 500 MHz

1.9.2 Mode broadening and photon lifetime

In the previous sub-section we considered the laser emission frequencies that are extremely sharp and have zero width. In practice, all the mode frequencies have finite width because of the reflection loss in the mirrors. We can find the time dependence of the light intensity in the cavity. We consider a passive optical resonator which is a laser cavity with no optical gain or amplification. The light inside the cavity gets reflected in the back and forth motion and loses energy in each optical refection. The light beam with intensity I at any point inside the cavity hits one wall and comes back with intensity $\rho_1 I$, where ρ_1 is the reflection coefficient. The same light beam then proceeds to the opposite wall and is reflected with the intensity $\rho_2\rho_1 I$, where ρ_2 is the reflection coefficient of the other wall. The change of intensity in one round-trip of distance $2L$ over a time $2L/c$ is

$$\delta I = (\rho_2\rho_1 - 1)I. \tag{1.9.4}$$

Hence the change of intensity per unit time is

$$\frac{\delta I}{\delta t} = \frac{(\rho_2\rho_1 - 1)I}{2L/c}. \tag{1.9.5}$$

The laser cavities usually have high reflectivity and the loss per reflection is small and the time interval δt is extremely small, so we can approximate

$$\frac{dI}{dt} = -\frac{1}{\tau}I.$$

where τ is the cavity lifetime or photon lifetime, so that

$$I(t) = I_0 e^{-t/\tau}. \tag{1.9.6}$$

Hence,

$$\tau = \frac{2L}{c(1-\rho_2\rho_1)} \tag{1.9.7}$$

is the photon lifetime defined as the time in which light intensity decays to $1/e$ of its initial value. This lifetime can be measured by sending a pulse into the cavity and measuring its intensity against time. This will lead to an exponential curve.

The frequency spectrum of the mode can be determined from its decay by using the time frequency relation. The electric field intensity is damped in this case and can be written as

$$\mathcal{E}(t) = \mathcal{E}_0 \, e^{-t/2\tau} \cos \omega t. \tag{1.9.8}$$

Thus the field is a damped oscillation as given in Fig. 1.23. It can be easily shown from Fourier analysis that this type of wave follows the uncertainty relation

$$\Delta\omega_{1/2}\tau \approx 1, \tag{1.9.9}$$

where $\Delta\omega_{1/2}$ is the angular frequency Full Width at Half Maximum (FWHM) and hence the frequency FWHM is

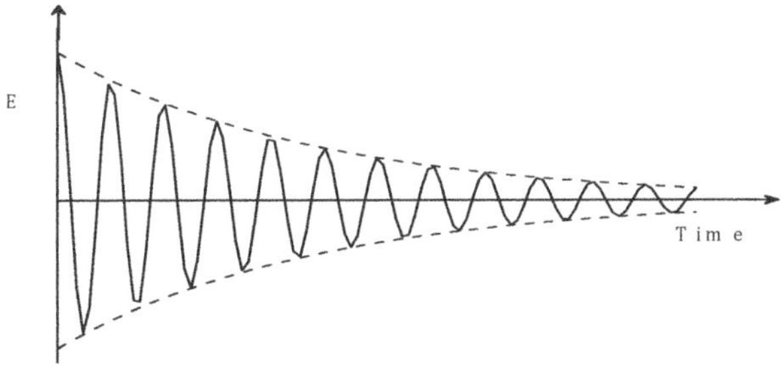

Fig. 1.23 Exponential decay of the damped oscillation.

$$\Delta\nu_{1/2} = \frac{1}{2\pi}\frac{c(1-\rho_2\rho_1)}{2L}. \tag{1.9.10}$$

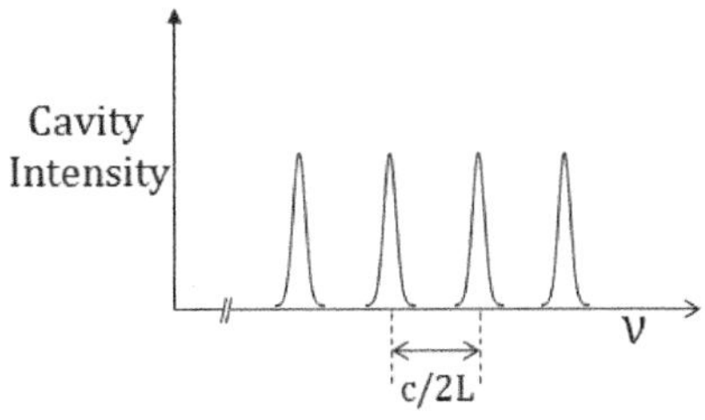

Fig. 1.24 Modes with a width of $\Delta\nu_{1/2}$ are separated by a distance of $\frac{c}{2L}$ in the frequency scale.

The frequency distribution will look similar to what is described in Fig. 1.24. The modes of width $\Delta\nu_{1/2}$ are separated by distance of $\frac{c}{2L}$. This separation is often referred to as Free Spectral Range (FSR). Each mode has width and they are discretely separated. This description is purely classical, but as we shall discuss in Chapter 6 these modes are the quanta of electromagnetic energy in a resonator. In that case the photons are the quanta of light. The frequency width can be related to time width in terms of the uncertainty relation between energy and

time
$\Delta(\hbar\omega_{1/2})\tau \approx \hbar$. Hence the uncertainty of photon energy is related to the photon life time. The coherence length is

$$L_c = \frac{c}{\Delta\nu_{1/2}} = \frac{4\pi L}{(1-\rho_2\rho_1)}. \tag{1.9.11}$$

Exercise 1.10 Find the frequency full width at half maximum of a laser with a resonator length of 30 cm and the walls having reflectivity of 0.95 and 0.7. Also calculate the cavity lifetime of the photon.

Solution

Frequency full width at half maximum,

$$\Delta\nu_{\frac{1}{2}} = \frac{1}{2\pi}\frac{c(1-\rho_2\rho_1)}{2L}$$

$$= \frac{1}{2\pi}\frac{3\times10^8(1-0.95\times0.7)}{2\times0.3} = 0.2667\times10^8 Hz = 26.67\ MHz$$

Cavity life time can be obtained from the above data,

$$\tau = \frac{2L}{c(1-\rho_2\rho_1)} = 5.97\times10^{-9}\ \text{sec} = 5.9\ nanosec$$

1.9.3 Quality Factor

The quality factor of a resonator is defined as the ratio of the centre frequency ν and the frequency width $\Delta\nu_{1/2}$

$$Q = \frac{\nu}{\Delta\nu_{1/2}} = \nu\frac{4\pi L}{c(1-\rho_2\rho_1)} \tag{1.9.12}$$

It is a measure of the quality of the resonator to sustain the wave with a narrow line width. Thus the relative sharpness of the modes is high for high reflectivity of the walls and large length of the resonator. For high reflectivity in the visible or near IR region thin dielectric layers are used. They are often better than metallic mirrors.

1.9.4 Cavity Finesse

With the increase of laser tube length the frequency half width decreases and the modes are sharper, but the frequency spacing of the modes also decreases. The cavity finesse is a measure of the ratio of the mode spacing and the mode width

$$F = \frac{c/2L}{\Delta\nu_{1/2}} = \frac{2\pi}{(1-\rho_2\rho_1)} \tag{1.9.13}$$

The finesse and the cavity quality factor are related by

$$Q = F\,\nu\,\frac{2L}{c} = Fn \tag{1.9.14}$$

Hence quality factor Q is related to cavity finesse F by the mode number n. All three quantities describe the spectral width of the cavity modes.

1.9.5 Gaussian Line Shape of Laser Beam

As described earlier the laser beam confined in a cavity has frequency modes with separation that depend on the resonator length. The beam propagates in a longitudinal direction along the length of the optical resonator. There is no propagation in the transverse direction. But the beam has a width in the transverse direction. It can be shown that the beam has a Gaussian shape and has therefore a natural transverse confinement and does not need any mirror for confinement of the laser beam in the transverse direction.

For a time dependent harmonic wave with frequency ω the Maxwell's equations lead to the wave equation in the form

$$(\nabla^2 + k^2)\mathcal{E}(x, y, z) = 0, \tag{1.9.15}$$

where $\mathcal{E}(x, y, z)$ is the complex electric field amplitude for any polarization direction of the electric field vector and $k = \omega/c$. This form of the equation will be clear in Chapter 6 when the wave equation of the electromagnetic field will be derived. As we noted earlier a laser like He-Ne has an optical confinement usually in a cylindrical form. The beam propagates along the axis of the cylinder; the radial direction gives the width of the beam. So we assume a cylindrical symmetry for the solution of the wave equation with $\mathcal{E}(x, y, z)$ replaced by $\mathcal{E}(r, \varphi, z)$. The wave equation becomes

$$\frac{\partial^2 \mathcal{E}}{\partial r^2} + \frac{1}{r}\frac{\partial \mathcal{E}}{\partial r} + \frac{\partial^2 \mathcal{E}}{\partial z^2} + k^2\mathcal{E} = 0, \tag{1.9.16}$$

where we have assumed that the electric field depends only on r in the transverse direction so that there is no dependence on φ. We consider only the magnitude of the electric field that is now not considered as a vector. In fact it is a vector in the plane perpendicular to the z-direction. If it has fixed polarization direction (say y), the electric field is along the x-direction.

Hence $\mathcal{E}(r, \phi, z)$ can be written as

$$\mathcal{E}(r, \phi, z) = \psi(r, z)e^{ikz}. \tag{1.9.17}$$

The oscillatory term takes into account the rapid spatial variation in space along the z-direction. The amplitude $\psi(r, z)$ gives the magnitude of the field in the transverse direction. It also varies with the propagation direction of the wave. Substituting into the wave equation (1.9.16) we can write

$$\frac{\partial^2 \psi}{\partial r^2} + \frac{1}{r}\frac{\partial \psi}{\partial r} + \frac{\partial^2 \psi}{\partial z^2} + 2ik\frac{\partial \psi}{\partial z} = 0. \tag{1.9.18}$$

Since ψ is a slowly varying function of z, we assume that $\frac{\partial^2 \psi}{\partial z^2}$ is much smaller than $\frac{\partial \psi}{\partial z}$ and hence can be omitted. So we get

$$\frac{\partial^2 \psi}{\partial r^2} + \frac{1}{r}\frac{\partial \psi}{\partial r} + 2ik\frac{\partial \psi}{\partial z} = 0. \tag{1.9.19}$$

The mathematical solution of the equation is still complicated and is worked out in some text-books like the book by Yariv. It is beyond the scope of the present book. But we shall discuss the possible solution and stress on the radial dependence of the field in the transverse direction. We can assume a trial solution in the form

$$\psi(r, z) = e^{-P(z)}e^{-ikr^2/2q(z)}, \tag{1.9.20}$$

where $P(z)$ and $q(z)$ are complex functions of z. It can be shown that we can get a solution in the form

$$\psi(r,z) = \psi_0 e^{iF(r,z)} e^{-r^2/w^2(z)} \tag{1.9.21}$$

$F(r,z)$ and w(z) are real functions., w(z) indicates spot size. $F(r,z)$ depends on r and z and it contains a term $R(z)$ that defines the diameter of the beam. The function $\psi(r,z)$ describes a wave that varies slowly in amplitude along the propagation direction (Fig.1.26). The amplitude decreases exponentially in the direction transverse to the propagation direction. Because of the Gaussian nature of the envelope of the field it is called a Gaussian beam. Following Eq. (1.9.17) and (1.9.21) ψ_0 is the maximum amplitude of the field at z=0 and r=0. It may be noted that the phase of the field defined by $F(r,z)$ depends on both the distance along the propagation direction and along the transverse direction.

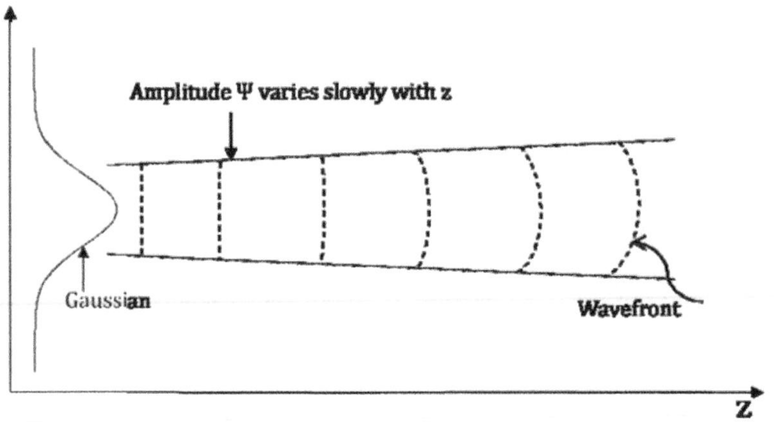

Fig. 1.25 Propagation of a Gaussian beam along z direction. At z=0, r=0, the laser has a minimum width and this is the beam waist.

The laser beam emanating from the output coupler or the pin-size hole of one mirror propagates along a straight line. At this point ($z = 0$) we can assume that the radius of curvature $R \to \infty$ or the wave front is planar. The width of the laser beam at this point may be called the beam waist $w_0 = w(z = 0)$. When the radius of curvature is infinity, the spot size has a minimum. With propagation of the beam along z, the spot size increases and the radius of curvature decreases so that the wave front is curved and the spot size is larger. But the shape remains Gaussian.

We consider the laser beam to be confined within two resonator mirrors of radii r_1 and r_2 (Fig. 1.26). A wave front that is diverging and moving towards one mirror will hit the mirror and will be reflected back as a converging wave. The beam will have a radius of curvature r_1. Similarly the other beam moving to the other mirror will have a radius of curvature r_2 . We consider a symmetric resonator with $r_1 = r_2 = r$.

The spot size and the front radius can be shown to vary with z as

$$w(z) = w_0[1 + \left(\frac{z}{z_0}\right)^2]^{1/2} \tag{1.9.22}$$

$$R(z) = z[1 + \left(\frac{z_0}{z}\right)^2] \tag{1.9.23}$$

For the Gaussian beam the electric field spatial distribution can be written as

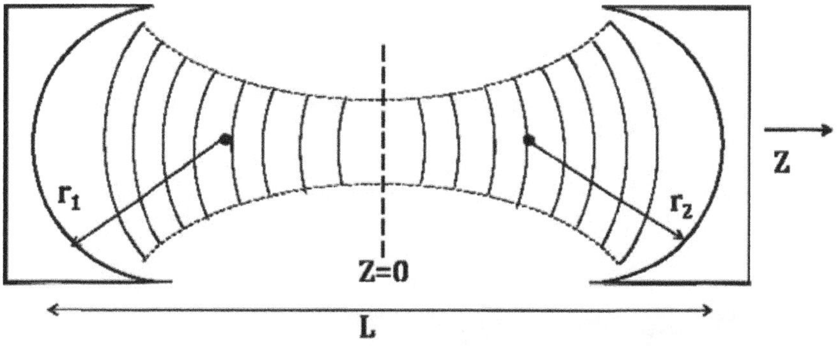

Fig. 1.26 Resonator of length L. The beam is confined between two mirrors of radii of curvature r_1 and r_2.

$$\mathcal{E}(r,z) = \mathcal{E}_0 \frac{w_0}{w(z)} e^{-r^2/w^2(z)}. \tag{1.9.24}$$

The parameter z_0 is defined as the Rayleigh length and is given by

$$z_0 = \frac{\pi {w_0}^2}{\lambda}. \tag{1.9.25}$$

The beam waist is located at the centre of the resonators at z=0, hence

$$r = R\left(\frac{L}{2}\right) = \left(\frac{L}{2}\right)\left[1 + \left(\frac{z_0}{L/2}\right)^2\right], \tag{1.9.26}$$

or,

$$z_0 = L/2\left(\frac{2r}{L} - 1\right)^{\frac{1}{2}}. \tag{1.9.27}$$

This leads to

$${w_0}^2 = \frac{\lambda L}{2\pi}\left(\frac{2r}{L} - 1\right)^{1/2}. \tag{1.9.28}$$

From the above results it may be noted that real values of spot size and Rayleigh length can be obtained only when $2r > L$, otherwise they cannot support a Gaussian wave. If $2r < L$, we cannot have a stable resonator. This is true for all cw lasers. This can be used for pulsed lasers in some special cases.

1.9.6 Higher order Laser Modes

We have so far considered one dimensional motion with a particular solution of Maxwell's equations. In fact, the solutions of Maxwell equations lead to Hermite-Gaussian modes. Gaussian wave is one of these modes (compare with the ground state solution of the linear harmonic oscillator). The Hermite Gaussian modes are the transverse electromagnetic (TEM) modes. The Gaussian modes are one of a special class of such modes, denoted as TEM_{00}. The modes are confined in three dimensions, hence we need three integers to define them. One of them corresponds to the propagation in the longitudinal direction and as considered earlier in Eq. (1.9.2), they are denoted by the integer n. We use the integers l and m to denote the transverse wave in the direction perpendicular to the propagation axis and the modes are denoted by TEM_{lm}. The electric field is given by the Hermite polynomial in the Hermite Gaussian form as

$$\mathcal{E}(x,y) = \mathcal{E}_0 H_l\left(\frac{\sqrt{2}}{w(z)}x\right) H_m\left(\frac{\sqrt{2}}{w(z)}y\right) e^{-\frac{x^2+y^2}{w(z)}}, \qquad (1.9.29)$$

H_l and H_m are the Hermite polynomials in x and y variables. The lowest order mode is the Gaussian, the next higher order mode is TEM_{10} and has the Gaussian envelope multiplied by x, giving a double peaked shape. For a mode of l-th order in the x-direction, there are l number of intensity zeroes and $l+1$ intensity maxima. Similar is the case for y-direction with m zeroes and $m+1$ maxima. We present sketches of a few lower order modes in Fig. 1.27.

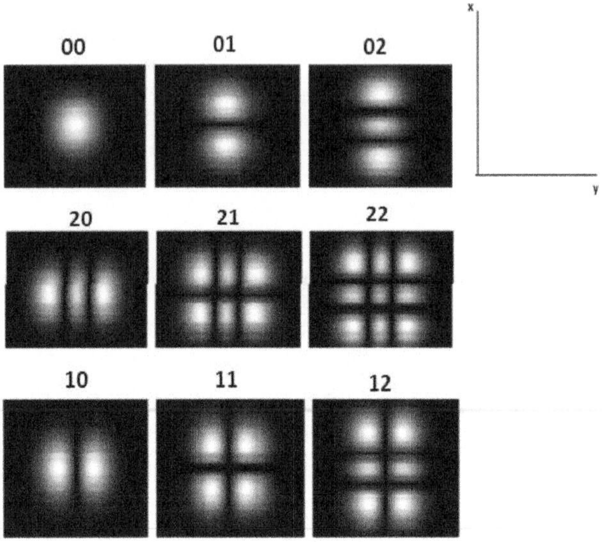

Fig. 1.27 Transverse Electromagnetic Mode TEM_{lm} showing the integers lm for the transverse directions x and y.

Higher order modes with different $l \neq 0$ and $m \neq 0$, give rise to higher order frequencies. If the mirrors are flat, these frequencies are degenerate and they all give the same frequency of $\nu = \frac{nc}{2L}$, as in the one dimensional case. The higher order modes with non-Gaussian shapes will still be present in this case. When the mirrors are slightly curved, the degeneracy is lifted and higher frequency modes will appear for higher order modes with show slight shift of frequencies.

1.10. Elementary Laser theory based on Rate Equations

We have noted above that the laser operation needs population inversion across the laser levels. The basic theory for understanding the laser action will be presented in Chapter 5, where we discuss the Semi classical Theory based on the interaction of classical electromagnetic field with the quantized atomic systems. In this section, we develop and solve the rate equations for a three-level atomic system to understand the population inversion and amplification process in a laser.

1.10.1 Three-level rate equation

For achieving population inversion between two levels we need a third level for transfer of atoms to the upper level. It can easily be shown that population inversion is not possible in a two-level system (Problem 1.20). We consider the atomic system with three non-degenerate energy levels E_1, E_2, and E_3 (Fig. 1.28). The population density of the respective levels are N_1, N_2, and N_3. A pump radiation with frequency ν_p is incident on the system in order to excite the atoms from level 1 to level 3. A radiation with frequency ν is also acting in resonance with the transition 1↔ 2. We shall consider all possible stimulated and spontaneous emission processes. The figure may be compared with the Fig. 1.3 where a pumping mechanism lifts the atoms from level 1 to level 3. In the presence of a pump radiation there will be both stimulated emission and absorption of photon; but the absorption will dominate because of much larger population in the lower level. The atoms from the upper level 3 will be quickly transferred to level 2, which have much longer life time. The transition from level 2 to level 1 is the laser transition. The emission from this level will dominate since there will be larger population in level 2 caused by population inversion created by transfer of atoms from level 1 to level 2 by the pumping mechanism. The upper level can be a broad level or a band of energy as in the case of solid state lasers.

The rate equations for the three levels describing the rate of change of population of the levels are

$$\frac{dN_1}{dt} = R_{31}N_3 + S_{31}(N_3 - N_1) + R_{21}N_2 + S_{21}(N_2 - N_1), \tag{1.10.1}$$

$$\frac{dN_2}{dt} = R_{32}N_3 - R_{21}N_2 + S_{21}(N_1 - N_2). \tag{1.10.2}$$

$$\frac{dN_3}{dt} = -R_{31}N_3 + S_{31}(N_1 - N_3) - R_{32}N_3 . \tag{1.10.3}$$

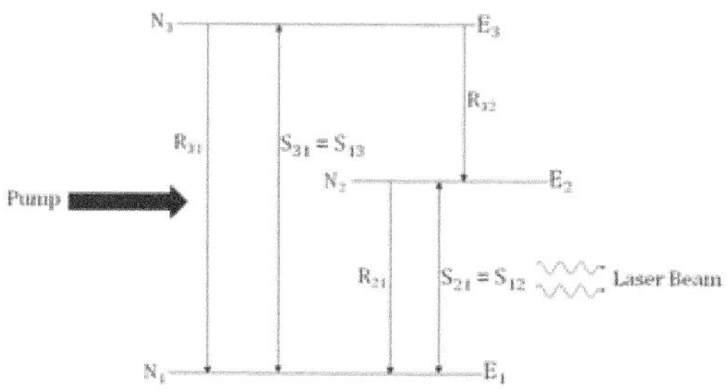

Fig. 1.28 Energy level diagram of a pumped three-level atom. Three spontaneous emissions from upper levels to lower levels are shown. Stimulated transitions across the pairs of levels 1,3 and 1,2 are also shown.

R_{ij} and S_{ij} represent the rate of spontaneous and stimulated transitions respectively across the levels i and j. R_{ij} include both radiative and non-radiative transitions. For laser transitions the radiative transitions will dominate. If the transitions are mostly radiative, then the constants R_{ij} are the same as the Einstein A-coefficients for spontaneous emissions across the levels i and j. The transition rates for stimulated processes may be related to the Einstein B-coefficients through the corresponding energy density as $S_{31} = B_{31}U_P(\omega)$ and $S_{21} = B_{21}U_L(\omega)$. The energy densities $U_P(\omega)$ and $U_L(\omega)$ are as defined in Sec 1.2 and correspond to the pump and laser transitions.

The terms $R_{31}N_3$ may be omitted since the atoms that reach the upper level 3 will make transitions mostly to the level 2 rather than to the level 1. The total number of atoms is

$$N = N_1 + N_2 + N_3 . \tag{1.10.4}$$

1.10.2 Population inversion in the steady state

At steady state, we should have

$$\frac{dN_1}{dt} = \frac{dN_2}{dt} = \frac{dN_3}{dt} = 0. \tag{1.10.5}$$

From Eq. (1.10.3) we can write

$$N_3 = \frac{S_{31}}{S_{31}+R_{32}} N_1. \tag{1.10.6}$$

From Eqs. (1.10.2) and (1.10.6) we get

$$N_2 = \frac{S_{31}R_{32}+S_{21}(S_{31}+R_{32})}{(S_{31}+R_{32})(S_{21}+R_{21})}. \tag{1.10.7}$$

From Eqs. (1.10.4), (1.10.6) and (1.10.7) we can write

$$\frac{N_2-N_1}{N} = \frac{S_{31}(R_{32}-R_{21})-R_{32}R_{21}}{3S_{31}S_{21}+2S_{31}R_{21}+2S_{21}R_{32}+R_{21}R_{32}+S_{31}R_{32}}. \tag{1.10.8}$$

From the above, it is clear that for population inversion, it is necessary that $N_2 - N_1$ should be positive. For this result to be true we need to satisfy the condition that

$$R_{32} > R_{21}.$$

Since the lifetime of the level is proportional to the inverse of R-coefficients, in order to satisfy the above criterion it is necessary that the level 2 has a longer life time compared to the level 3. If this condition is satisfied, we also need from Eq (1.10.8) that the pumping rate S_{31} should have a minimum value

$$S_{31}^{th} = \frac{R_{32}R_{21}}{(R_{32}-R_{21})}. \tag{1.10.9}$$

If $R_{32} \gg R_{21}$,

$$\frac{N_2-N_1}{N} = \frac{R_{32}(S_{31}-R_{21})}{3S_{31}S_{21}+2S_{21}R_{32}+R_{32}(S_{31}+R_{21})}. \tag{1.10.10}$$

If we work below laser threshold, S_{21} is very small, then we can get

$$\frac{N_2-N_1}{N} = \frac{S_{31}-R_{21}}{S_{31}+R_{21}}. \tag{1.10.11}$$

Hence, below threshold the population difference is independent of the transition rate S_{21} or the radiation intensity. Thus the population difference is large and the amplification continues to increase. As a result, the radiation intensity increases, and so the population difference decreases till it attains the threshold value. Thus, when the laser starts oscillating the population difference $N_2 - N_1$ remains close to the steady state value. We shall discuss the steady state operation in more details, when we describe semi-classical laser theory in Chapter 5.

When R_{32} is large, N_3 becomes very small, since all the atoms are drained to the level 2. Hence we can write $N_2 + N_1 \cong N$. From Eq. (1.10.11) we can write

$$\frac{N_2-N_1}{N_2+N_1} = \frac{S_{31}-R_{21}}{S_{31}+R_{21}}, \tag{1.10.12}$$

or,

$$N_2R_{21} = N_1S_{31}. \tag{1.10.13}$$

Thus the rate of transfer of atoms per unit volume from level 1 to level 2 via the pumping through level 3, N_1S_{31} should be equal to the rate of spontaneous emission from level 2 to level 1 under steady state condition operating below threshold. W know that for usual laser operation the population difference $N_2 - N_1$ or population inversion above threshold is much small compared to the total number of atoms N. This allows us to assume that $N_2 \cong N_1$. Hence, the threshold value of S_{31} required to pump the laser is nearly the same as the spontaneous emission rate R_{21}, or $S_{31} \cong R_{21}$. If ν_p is the pump frequency to the level 3, then the power required per unit volume is

$$P = N_1 S_{31} \nu_p. \quad (1.10.14)$$

Hence, the threshold pump power is

$$P^{th} = N_1 S_{31}^{th} h\nu_p \cong N_1 R_{21} h\nu_p. \quad (1.10.15)$$

From the above assumption on level populations we can write $N_2 \sim N_1 = N/2$. Hence

$$P^{th} = \frac{NR_{21}h\nu_p}{2} = \frac{Nh\nu_p}{2\tau_{sp}}, \quad (1.10.16)$$

where we assumed R_{21} is the inverse of spontaneous emission life time.

Exercise 1.11 Calculate the threshold pump power for a He-Ne gas laser, the pump frequency, $\nu_p \sim 10^{15} Hz$, with spontaneous life time $\tau_{sp} \sim 10^{-8}\ sec$ and number density of $N \sim 1.6\ x\ 10^{19} per\ c.c.$

Solution

In this problem the pump frequency is close to the laser operation frequency. We have

$$\nu_p \sim 10^{15} Hz, \quad \tau_{sp} \sim 10^{-8}\ sec \text{ and } N \sim 1.6\ x\ 10^{19} per\ c.c.$$

Hence the threshold pump power is

$$P = \frac{1.6\ x\ 10^{19} \times 6.62 \times 10^{-27} \times 10^{15}}{2 \times 10^{-8}} erg\ per\ sec\ per\ c.c.$$

$$\cong 10^9 Watt\ per\ c.c.$$

1.11 Operation of Pulse Laser

Pulsed sources are useful for many applications. The pulsed operation can, in principle, be obtained from a continuous wave laser. But pulsed laser operation can be achieved more efficiently by controlling the population inversion energy built up in the cavity. There are two kinds of such operations:- (1) Q-switching and (2) mode locking technique. In a Q-switched laser operation, high energy laser pulses may be obtained for nanosecond duration. In the Mode Locking technique, low energy pulses of extremely short duration of the order of femtosecond can be obtained.

1.11.1 Q-switching

In this process, a shutter is placed inside the cavity such that it can be opened and closed intermittently. When the shutter is closed and the pumping process is on, there will be no feedback as the photons are prevented by the shutter to be reflected back by the mirror into the cavity. But population inversion will continue to increase and it will go beyond the threshold value for laser operation in absence of the shutter. The population is transferred from the ground state to the excited state, but there is no stimulated emission in absence of the feedback. After a substantial population inversion beyond threshold value is achieved, the shutter may be opened. The stimulated emission will start as the spontaneously emitted photons are reflected back into the cavity. Because of the large population build-up in the excited state the intensity of the emitted radiation will be very large. This will continue for a short duration as the population inversion will be consumed in the process. As soon as the shutter is opened, the Quality Factor Q increases suddenly. This phenomenon is called Q-switching. For high Q, the loss will be small. As the loss will start gaining, Q will become lower. At this point, the shutter may be closed again to allow the population inversion to build up within the resonator again. There will be no emission, which may start again as soon as the shutter is suddenly opened. This process may be continued, so that pulses of high energy are emitted at regular intervals. In a typical Nd-YAG laser the pulse interval is 0.1 sec, the energy of the order of one Joule is obtained in a pulse of ns duration. This will lead to an intensity of one gegawatt.

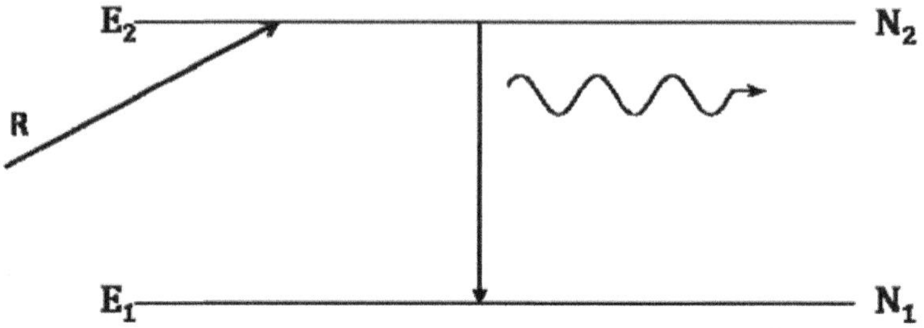

Fig.1.29 Two-level atom in the presence of a pump

In order to discuss the population build up, the energy and time duration of pulses in Q-switching we consider the two level atomic system (Fig. 1.29) that are involved in the laser emission process. The excited level gets populated by the pumping process as discussed earlier through another level or an electrical discharge etc. We designate the pumping rate by R. There will be both stimulated and spontaneous emission from the upper level. The number of stimulated emissions per unit volume is proportional to the number of photons n, it is designated by the rate Sn/V where S is the rate of stimulated emission. The spontaneous emission out of the level 2 may be neglected during the short duration of the pulse, although spontaneous emission is necessary to initiate the process. So we can write from Eq. (1.10.2)

$$\frac{dN_2}{dt} = R + \frac{Sn}{V}(N_1 - N_2). \tag{1.11.1}$$

During the short pulse duration we can omit the pumping rate, it is necessary to build up the population inversion at atime before the pulse is initiated. The following treatment discusses the behaviour of population inversion and photon number change during the short pulse duration only.

$$\frac{dN_2}{dt} = -\frac{Sn}{V}\Delta N\,, \tag{1.11.2}$$

where,

$$\Delta N = N_2 - N_1. \tag{1.11.3}$$

Similarly, we can write rate equation for the lower level population

$$\frac{dN_1}{dt} = \frac{Sn}{V}\Delta N. \tag{1.11.4}$$

From the above, we can write

$$\frac{d\Delta N}{dt} = \frac{-2Sn}{V}\Delta N. \quad (1.11.5)$$

The rate of change of photon number can be written as

$$\frac{dn}{dt} = Sn\Delta N - n/\tau_s\,, \quad (1.11.6)$$

τ_s is the exponential decay rate from the upper state. The spontaneous emission is omitted. The threshold population inversion, when $\frac{dn}{dt} = 0$, can be written as

$$\Delta N_{th} = 1/S\tau_s \quad (1.11.7)$$

Hence,

$$\frac{dn}{dt} = \frac{n}{\tau_s}[\frac{\Delta N}{\Delta N_{th}} - 1]. \quad (1.11.8)$$

At the beginning of the pulse, the population inversion is above threshold, so $\Delta N > \Delta N_{th}$, the photon number will increase. This will continue till $\Delta N = \Delta N_{th}$, after that the number of photons will go on decreasing. This describes the Q-switching process when the Quality suddenly jumps to a high value and falls again. The energy of the pulse will pick up and decay in this short duration.

The variation of photon number with population inversion can be written from Eq. (1.11.5), (1.11.7) and (1.11.8) as

$$\frac{dn}{d\Delta N} = \frac{V}{2}[\frac{\Delta N_{th}}{\Delta N} - 1] \quad (1.11.9)$$

Integrating this equation we can get the total photon number with respect to an initial number n_i of photons in terms of the population inversion at threshold and also the population inversion at the initial and final time of the pulse.

In order to calculate the total energy we can calculate the total number of photons generated in the process. We obtain from Eq. (1.11.5), (1.11.6)

$$\frac{dn}{dt} = -\frac{V}{2}\frac{d\Delta N}{dt} - n/\tau_s. \quad (1.11.10)$$

Integrating this equation from t=0 to t=∞ we get

$$n_f - n_i = \frac{V}{2}[\Delta N_i - \Delta N_f] - \frac{1}{\tau_s}\int_0^\infty n dt.$$

The initial and final photon numbers are small compared to the total integrated number of photons, hence we can neglect the term, $n_f - n_i$. Hence, we get

$$\int_0^\infty n dt = \frac{V\tau_s}{2}[\Delta N_i - \Delta N_f]. \quad (1.11.11)$$

The total output power is

$$P = \frac{nh\nu}{\tau_s}. \quad (1.11.12)$$

Hence the total energy of the pulse is

$$E = \int_0^\infty P dt = \frac{h\nu}{\tau_s}\int_0^\infty n dt$$

$$= \frac{V}{2}\left[\Delta N_i - \Delta N_f\right] h\nu. \qquad (1.11.13)$$

It is evident that when the population inversion changes from ΔN_i to ΔN_f, there is change of photon number by $\frac{1}{2}\left[\Delta N_i - \Delta N_f\right]$.

The pulse duration can be obtained from the total energy and the maximum power as

$$\tau_d = \frac{E}{P_{max}} = \frac{V\tau_s}{2}\left[\Delta N_i - \Delta N_f\right]/n_{max} \qquad (1.11.14)$$

n_{max} can be obtained by integration of Eq. (1.11.9) expressed in terms of population inversion and its threshold value.

Experimental methods of Q-switching

Several methods have been used to achieve Q-switching in a laser. The most common and frequently used methods are : (1) Electro-optic switches, (2) Rotating Prisms, (3) Acoust-optic Switches and (4) Saturable Absorbers. There are two different types of switches, active and passive. The electro-optic switches are active switches in which an external operation like change of the applied voltage to electro-optic shutters is used. Saturable absorbers are passive elements, where switching operation is automatically achieved because of the nonlinearity of the medium. We shall discuss electro-optic shutters and saturable absorbers. In the case of mechanical devices employing rotation around an axis of the resonator, the rotation speed cannot be very high, so instantaneous switching from low to high Q value is difficult. Electronically controlled shutters are very efficient.

Electro-optic switch

In this kind of switch electro-optical effect or Pockels effect is used. The Pockels cell (based on Pockels effect) consists of a crystal of potassium dihydrogen phosphate (KDP) or lithium niobate for visible or near infrared region or a cadmium telluride crystal for mid-infrared region. A dc electric voltage is applied to the crystal and it changes the refractive indices of the crystal. The birefringence induced in the crystal is proportional to the applied voltage. The system contains the electro-optic cell and a polarizer (Fig. 1.30).

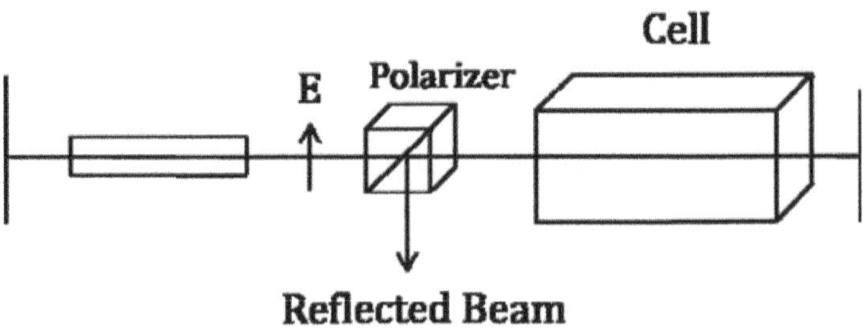

Fig. 1.30 Electro-optic cell

The cell is oriented and biased in such a way that the x and y axes of birefringence are in the plane orthogonal to the axis of the resonator. The polarizer axis is at an angle of 45° to the birefringence axes. We consider the laser beam, originating from the active medium in the resonator with a polarization parallel to the axis of the polarizer, to pass through the combination of the polarizer and the cell. It will be transmitted through the polarizer and will be incident on the cell. In the cell, the birefringence axes are at $\pi/4$ to the electric field of the radiation. The electric field will be resolved into two components E_x and E_y, oscillating in phase (Fig. 1.31(a)). After passage through the cell, the two component beams will experience different phase changes, so there will be a phase difference induced in the two components.

The applied voltage may be adjusted such that the phase difference introduced is $\pi/2$. Hence the two components coming out of the cell will have a phase difference such that when the x-component is a maximum, the y component will be a minimum, and *vice versa.*

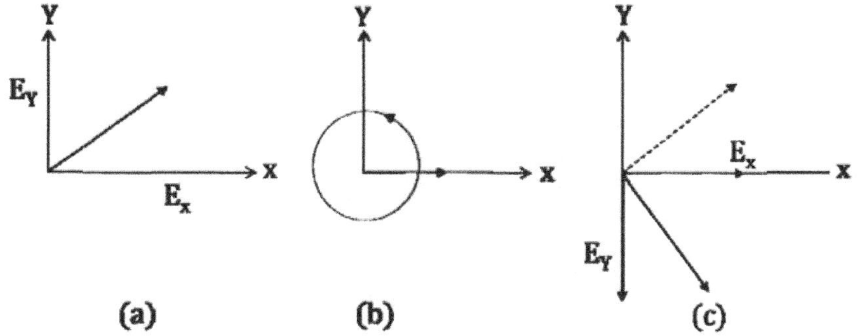

Fig. 1.31 (a) Incident wave with the resolved electric field components, (b) the circularly polarized wave after passage through the cell, (c) the reflected wave with a total phase change of π.

In this case, the cell functions as a quarter wave plate. The wave will be circularly polarized (Fig. 1.31(b)). If it is circularly polarized in the clockwise direction, after reflection from the mirror, it will be circularly polarized in the counter clockwise direction. This wave, after passing through the cell, will again suffer a phase change of $\pi/2$. So there is a total phase change between the two components of π. The wave will now be linearly polarized, but if it has a positive maximum for E_x, the E_y component will have a negative maximum. The net field will be polarized at an angle of 90° to the original beam (Fig. 1.31(c)). Hence it will not pass through the polarizer and will be reflected out of the resonator system. This is a condition for Q-switching. As soon as the voltage applied is switched off, the birefringence will disappear and the phase difference of the components will be zero. Hence the beam will pass through the cell and the polarizer and reflected back by the mirror into the resonator. The alternate switching off and on of the applied voltage can induce change of Q value of the resonator producing pulses. Thus Q switching is achieved by applying a dc voltage to the Electro-optic shutter and removing it when the highest inversion is achieved. The voltage required is such that the phase change is

$$\Delta\varphi = \frac{\pi}{2}.$$

But the phase change caused by birefringence is

$$\Delta\varphi = \frac{2\pi}{\lambda}\Delta nL,$$

where Δn is the change of refractive index and L is the length of the cavity and λ is the wave length of the radiation.

Hence,

$$\Delta nL = \frac{\lambda}{4} \quad (1.11.15)$$

The difference in optical path length of the two polarizations is equal to $\frac{\lambda}{4}$. Hence the cell acts like a quarter wave plate and the voltage is called the quarter wave voltage.

Saturable absorber

In this technique, a saturable absorber (e.g. a dye dissolved in an organic solvent and placed in a cell) is kept inside the cavity between the laser resonator and the mirror (Fig. 1.32). The absorption coefficient of the medium is decreased with an increase in the intensity of the incident radiation, since the transition gets saturated. At the beginning, the unsaturated absorption is low. The pumping action starts to build up population inversion, but laser intensity will be very low as the absorber does not allow feedback. The laser action will start only when the gain of the active medium compensates the losses due to the absorber and other losses in the cavity. Eventually, the laser action starts and with the increase of the radiation intensity, the absorber starts to bleach owing to saturation and the intensity of laser radiation increases sharply. This will lead to a sharp increase in Q and result in a strong pulse. The energy stored in the medium will produce a pulse of short duration. This process will continue when the absorption starts again.

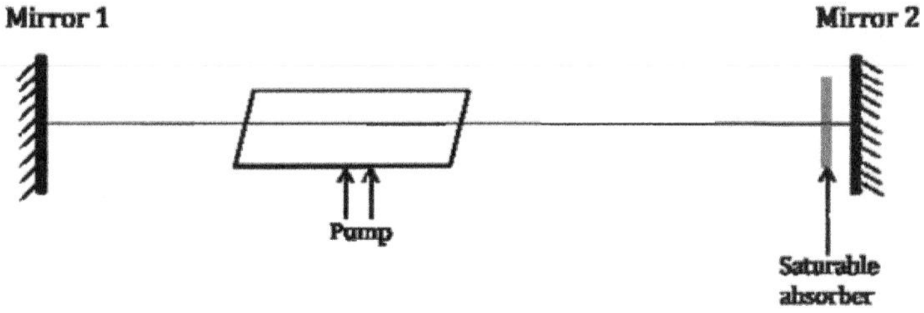

Fig. 1.32 Saturable absorber for mode locking.

1.11.2 Mode Locking

Mode locking technique is used to produce ultra-short pulses of duration picosecond or smaller than that. In order to understand the mode locking process we consider beat formation in the case of interference of two harmonic waves. If two waves, each with constant intensity and slightly different frequencies, are superposed, they can interfere constructively when they are in phase resulting in a maximum intensity. But if they are out of phase, they will interfere destructively and the resultant intensity may be zero (Fig.1.33). This periodic change of intensity produces beat frequencies. When there are more than two waves they will beat with each other resulting in a complex wave pattern.

We consider a laser beam having many close-lying frequencies. In the case of a laser oscillating simultaneously in many modes, there is no correlation or fixed phase relationship between the modes. A superposition of these waves will produce a random wave pattern that will look like a noise. If we can lock the phases of each of these modes, or bring them all in the same phase at any time and maintain this phase relationship, depending on the number of such waves and their frequency spacing, at some intervals of time they will have all their maxima put together and similarly at some other time they will have all their minima placed together.

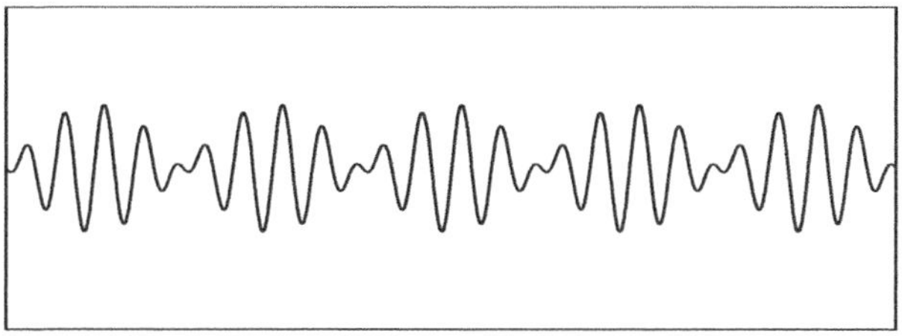

Fig. 1.33 Beat frequencies showing periodic change of intensity.

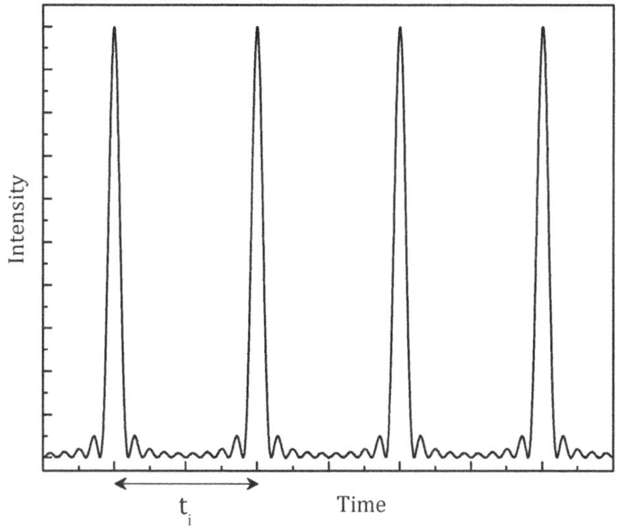

Fig. 1.34 Intensity variation of a mode locked laser.

In this case the output of the laser will look like a series of repetitive pulses and these pulses are mode locked and the method is called mode locking. The intensity variation of a mode locked laser (Fig. 1.34) shows periodic pulses of higher intensity than the intensity of the random pulses. The pulse width is inversely proportional to the number of modes used for mode locking.

In a laser with the mirrors having a distance l and the frequency band width of the laser radiation defined by the gain profile of the active medium Δv, the intermodal spacing will be $\delta v = c/2nl$, the number of modes within the bandwidth will be large and determined by

$$N = 2nl\,\Delta v/c, \tag{1.11.16}$$

where the integer value closest to and larger than the value of the term is to be taken. In a typical case, the band width is of the order of 50 GHz, the length is around 50 cm, so the value of N can be 167 (if n=1). All the modes oscillate independently and with random phase distribution in the range $-\pi$ to $+\pi$. The total intensity of all such modes in the profile can be written as

$$I = C|\mathcal{E}(t)|^2 = C\left|\sum_{-N/2}^{+N/2} \mathcal{E}_{0q} \exp\left(2\pi i \nu_q t + \varphi_q\right)\right|^2, \quad (1.11.17)$$

where

$$\nu_q = \nu_c + q\delta\nu,$$

q varies from $-N/2$ to $+N/2$ in integer numbers. ν_c is the centre frequency. Hence we can write

$$I = C\left|\sum_{-N/2}^{+N/2} \mathcal{E}_{0q} \exp\left(2\pi i q\delta\nu t + \varphi_q\right)\right|^2$$

$$I = C\sum\left|\mathcal{E}_{0q}\right|^2 + C\sum_q\sum_r \mathcal{E}_{0r}{}^{*}\mathcal{E}_{0q}\exp[2\pi i(q-r)\,\delta\nu t + i(\varphi_q - \varphi_r)]. \quad (1.11.18)$$

The phase varies at random in the region from $-\pi$ to $+\pi$, if the number of modes is large, the second term on the rught hand side will be small. The intensity, determined by the first term, is a random superposition of beam intensities of the various modes. When the system is operating below threshold, the modes are uncorrelated; the spontaneously emitted photons are not correlated. When the system is above threshold, the fluctuations are less, but the modes still remain uncorrelated.

If we set the phase of all the modes to a particular fixed value, say φ_0, they are phase-locked at some time t=0, in that case, the intensity can be written as

$$I = C\left|\sum_{-N/2}^{+N/2} \mathcal{E}_{0q} \exp\left(2\pi i q\delta\nu t\right)\right|^2. \quad (1.11.19)$$

If all the modes have the same amplitude, $\mathcal{E}_{0q} = \mathcal{E}_0$, then the intensity is

$$I = I_0\left|\sum_{-N/2}^{+N/2} \exp\left(2\pi i q\delta\nu t\right)\right|^2 \quad (1.11.20)$$

Where, $I_0 = C|\mathcal{E}_0|^2$, the summation can be calculated to get

$$I = I_0\left[\frac{\sin \pi(N+1)\delta\nu t)}{\sin(\pi\delta\nu t)}\right]^2 \quad (1.11.21)$$

The time variation of intensity is shown in Fig. 1.34. It shows a shape similar to that of a diffraction grating, There are series of pulses at a definite spacing determined by $\delta\nu$. The pulses are separated by a time duration

$$t_i = \frac{1}{\delta\nu} = \frac{2nl}{c}. \quad (1.11.22)$$

This time is the same as time taken by the beam passage through the cavity back and forth. It appears that after one round trip of the beam through the cavity there is a pulse of light output. It is found that the intensity falls off very rapidly on either side of the peak, since N is very large. The time difference between the two instants when the intensity is zero on either side of the peak is

$$\Delta t = \frac{2}{(N+1)\delta\nu} = \frac{2}{\Delta\nu}. \quad (1.11.23)$$

Hence the pulse duration is approximately equal to the inverse of the band width. Thus, the larger the bandwidth, the smaller is the pulse width, In the case of dye laser, the band width is large (nearly 10000 GHz), the pulse width is 0.1 psec, whereas in the case of He-Ne laser the band width is 1.5 GHz and the pulse width is 670 psec. The peak intensity of each mode-locked pulse is

$$I = (N+1)^2 I_0. \tag{1.11.24}$$

Thus the peak intensity of each pulse is $(N+1)$ times larger than the unlocked laser intensity. Hence there is a large enhancement of peak intensity for a large value of N.

Methods of Mode Locking

For effective mode locking we need to couple various modes with each other. It can be achieved by active or passive methods. In the active method, an external modulation of the optical path length is used. In the passive method, a saturable absorber is placed inside the cavity.

A loss modulator with a modulation frequency equal to the intermodal frequency spacing $\delta\nu$ is placed inside the resonator cavity. The loss of the cavity is modulated at this frequency, hence the amplitude of the field will also be modulated at the frequency $\delta\nu = c/2nL$. Hence the amplitude modulated wave for a particular frequency may be written as

$$(I_0 + I_{0m}\cos 2\pi\,\delta\nu t)\cos 2\pi\nu_m t = I_0\cos 2\pi\nu_m t + \frac{1}{2}I_{0m}\cos 2\pi\,(\nu_m + \delta\nu)t + \frac{1}{2}I_{0m}\cos 2\pi\,(\nu_m - \delta\nu)t. \tag{1.11.25}$$

Hence the amplitude modulated wave at frequency ν_m generates two sideband frequencies at $\nu_m + \delta\nu$ and $\nu_m - \delta\nu$. Thus we have three frequencies oscillating at frequencies ν_m, $\nu_m + \delta\nu$ and $\nu_m - \delta\nu$. They are separated by the intermodal frequency. The side band frequencies at $\nu_m \pm \delta\nu$ induce all three frequencies to oscillate with a perfect phase relationship among all three modes. The amplitude of these frequencies are also modulated at the frequency $\delta\nu$. So this modulation will generate two new sidebands at $\nu_m \pm 2\delta\nu$. They will again have perfect phase correlation among all the five modes, and so on. Thus all the modes have to follow a perfect phase correlation and this will lead to mode locking.

Acousto-optic modulators are also used for active mode locking method. For passive modulation saturable absorbers are used.

1.12 Non-linear effects in Laser Physics

1.12.1 Nonlinear optical effects

In a linear process, two light waves of different frequency interacting with a material medium do not interact with each other. They propagate independently through the medium. The presence of the second wave does not affect the first wave in any way. But in a non-linear medium these waves are no longer uncoupled. The non-linear interaction within the medium can alter the amplitude and the phase of the wave. As a result of the interaction, there can be generation of new waves with frequency different from those of the incident waves. Such frequency conversion is not possible in a linear medium. If the medium is a perfect vacuum, there can be no coupling, because Maxwell's equations are linear in electric and magnetic fields. Hence superposition principle states that the sum of the two solutions will also be a solution of the same equations. The coupling of the waves is a physical process in which one of the waves interact with the medium and can alter the properties of the medium, such that the wave interacts with the altered medium and so its properties are changed. The degree of the coupling depends on how strongly the wave interacts with the medium.

For normal light beam the light-atom interaction is weak. If the electric field associated with an ordinary source of light is of the order 500 V/m, the same for the atom field is 10^{11} V/m. Hence the light field is so weak that it cannot cause any effect on the electron motion in the atom. Hence the medium remains unperturbed. This is an advantage for use of light beam for many purposes, for example, in the case of imaging. Since the response from the lens is linear there is no distortion in the image.

But this scenario changes with the use of laser radiation that has much higher light intensity. Because of the higher intensity, the light wave interacts with the material such that some new properties develop. When a laser beam with frequency ω is incident on the material medium, a wave with a new frequency 2ω may be generated. This phenomenon is called Second Harmonic Generation (SHG). It was discovered in 1961, soon after the successful operation of the first laser . If a red color beam of wavelength 750 nm falls on a non-linear crystal, a violet color of light at wavelength 375 nm will emerge. Similarly, there can be sum and difference frequency generation. If the light beams with frequency $\omega_1$and ω_2 are incident on the non-linear medium, there can be generation of two new waves with the sum frequency $\omega_1 + \omega_2$ and the difference frequency $\omega_1 - \omega_2$. In a parametric fluorescence process, a light beam with frequency ω spontaneously splits into two modes with frequencies ω_1 and ω_2. The energy conservation demands that $\omega = \omega_1 + \omega_2$. This phenomenon of Optical Parametric Oscillation (OPO) was discovered in 1965.

Since the discovery of laser, a new subject of study called Non-linear Optics has developed. The subject matter is very extensive and has been used in a large number of cases. We cannot describe it fully. We shall consider two aspects, optical parametric oscillation and second harmonic generation.

1.12.2 Non-linear polarization

In a linear medium we assume that the polarization in a medium in the presence of an electric field $\mathcal{E}$ is proportional to the field, so

$$P = \epsilon_0 \chi \, \mathcal{E}, \tag{1.12.1}$$

where ϵ_0 is the permittivity of the medium and χ is the linear dielectric susceptibility. In a non-linear medium this relation gets modified and we have to include the higher powers of the electric field.

$$P = \epsilon_0 [\chi \, \mathcal{E} + \chi' \mathcal{E}^2 + \chi'' \mathcal{E}^3 + \cdots], \tag{1.12.2}$$

χ', χ'' are higher order non-linear susceptibility. The second term in the above gives rise to SHG, sum and difference frequency generation, parametric oscillation etc. The third term leads to the third harmonic generation, four wave mixing, self phase modulation etc. For crystals with inversion symmetry the second order non-linear term vanishes. This is so because in these crystals, if the sign of the electric field is reversed, the sign of the polarization should also reverse. The term proportional to $\mathcal{E}^2$ does not change sign on reversal of sign of $\mathcal{E}$. Hence, we must have, $\chi' = 0$. Silica used in optical fibre communication has inversion symmetry. Hence they do not exhibit second order non-linear effect. Second order effect is found in crystals of lithium niobate $LiNbO_3$, potassium dihydrogen phosphate KH_2PO_4 etc. They can exhibit OPO behavior.

1.12.3 Second Harmonic Generation

By sending high power laser beam through crystal like quartz, we can generate a second harmonic frequency in addition to the original incident frequency. It was first demonstrated by using a Ruby laser beam. The laser beam is focused by a lens into a crystal that converts a part of the light energy with wavelength λ in the visible region into a wave of wavelength $\lambda/2$ in the ultraviolet region (Fig. 1.35). The output of the crystal can be analyzed by a spectrophotometer to observe the second harmonic wave.

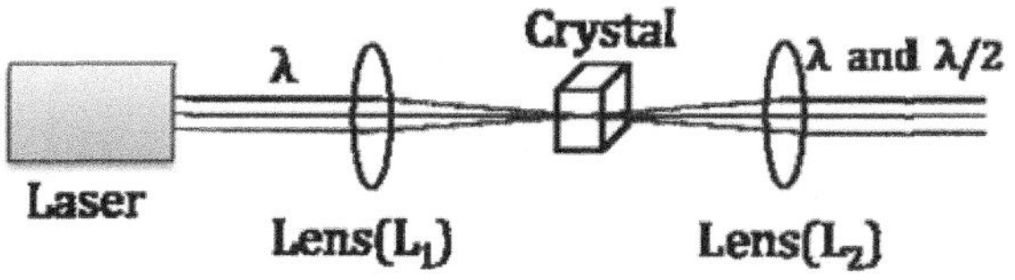

Fig. 1.35 Laser radiation focused by a lens on the crystal, the output beam shows an additional wave with frequency twice the original frequency.

A monochromatic plane wave with frequency ω is considered as propagating with uniform intensity within the crystal along a certain direction in space (say, z). We assume the origin of the z-axis to be at the entrance point of the beam at the crystal surface. The electric field of the electromagnetic wave may be written as

$$\mathcal{E}(z,t) = \frac{1}{2}[\mathcal{E}_0(z,\omega)\exp\{i(\omega t - kz)\} + c.c.], \tag{1.12.3}$$

c.c. denotes the complex conjugate of the first term in the bracket. $\mathcal{E}_0(z,\omega)$ is the complex amplitude of the electric field. The component of the wave propagation vector along z direction is

$$k = \frac{\omega}{v_p} = \frac{n\omega}{c}, \tag{1.12.4}$$

where v_p is the phase velocity of the light within the medium, n is the refractive index of the medium at frequency ω. Substituting the electric field, given by Eq. (1.12.3), in the second term of the total polarization given by Eq. (1.12.2) we find that there is one term in the second order polarization in the form

$$P'(z,t) = \frac{1}{2}\epsilon_0\chi'[\mathcal{E}_0^2(z,\omega)\exp\{i(2\omega t - 2kz)\} + c.c.]. \tag{1.12.5}$$

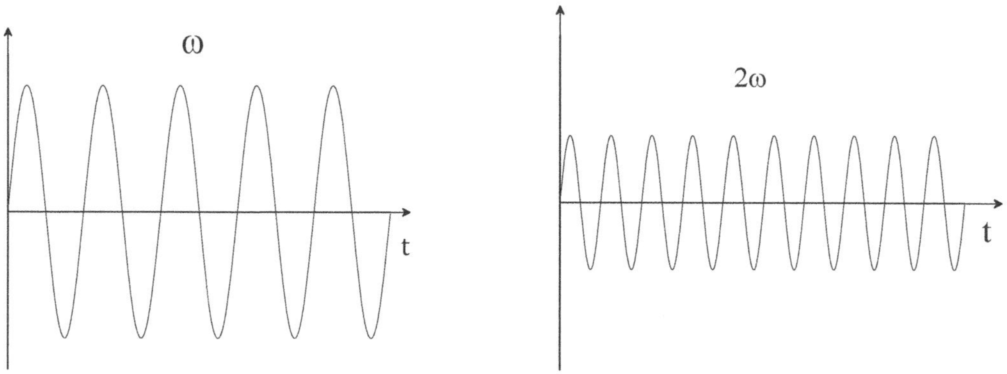

Fig. 1.36 Incident radiation at frequency ω and the second harmonic wave at frequency 2ω.

This term shows the polarization as a wave oscillating at frequency 2ω and propagation constant$2k$. This wave can radiate at a frequency 2ω. Thus the wave with second harmonic frequency may be created. The radiated electric field can be written as

$$\mathcal{E}_{SH}(z,t) = \frac{1}{2}[\mathcal{E}_{0SH}(z, 2\omega)\exp\{i(2\omega t - k_2 z)\} + c.c.]. \quad (1.12.6)$$

The non-linear effects are clearly shown in the Fig. 1.36 as the waves with a certain frequency and its second harmonic.

This wave has a propagation constant

$$k_2 = \frac{2\omega}{v_p} = \frac{2n\omega}{c} = 2k. \quad (1.12.7)$$

We assume that the refractive index for the second harmonic remains the same as that of the original wave. For this condition to be satisfied, the phase velocity of the polarization wave $\frac{2\omega}{2k}$ should be the same as the phase velocity of the generated wave $\frac{2\omega}{k_2}$. This condition is known as phase matching condition. If this condition is not satisfied the phase of the polarization wave will have a lag with the phase of the generated wave. This will not help the buildup of the SHG. Physical explanation is that the electromagnetic wave at frequency ω will beat with itself to produce a second harmonic wave at frequency 2ω. In case there is difference of refractive indices with frequency, there will be dispersion $\Delta n = n_{SH} - n$. This will cause a limit to the length l_c over which the phase of the polarization wave can match the phase of the second harmonic wave. We can define the coherence length by

$$k_2 l_c - 2k l_c = \pi \quad (1.12.8)$$

This is zero in the phase matching condition. But this may remain valid for a short distance. From Eq. (1.12.4) and (1.12.8) we can show

$$l_c = \frac{\pi}{k_2 - 2k} = \frac{\pi c}{2\omega \Delta n} \quad (1.12.9)$$

Hence a passage of the beam through the above distance may cause the waves to be out of phase by 180° and after passing through this distance the SH wave becomes weaker in intensity.

Exercise 1.12 Calculate the coherence length for a wave at wavelength 600 nm, so that the passage of the beam through this distance may cause a change of phase by π, if the change of refractive index is 0.01.

Solution

Coherence length $l_c = \frac{\pi c}{2\omega\Delta n}$,

in this problem, $\lambda = 600\ nm$, $\Delta n = 0.01$

Since, $\omega = \frac{2\pi c}{\lambda}$, $l_c = \frac{\pi c}{2\omega \Delta n} = \frac{\lambda}{4\Delta n} = 15\ \mu m$

1.12.4 Theoretical treatment

For theoretical explanation of the generation of SH waves we can consider the Maxwell equations.

$$\nabla \times \mathcal{H} = J + \frac{\partial D}{\partial t}, \quad (1.12.10)$$

$$\nabla \times \mathcal{E} = -\frac{\partial \boldsymbol{B}}{\partial t}, \tag{1.12.11}$$

$$\nabla . \boldsymbol{D} = \rho, \tag{1.12.12}$$

$$\nabla . \boldsymbol{B} = 0. \tag{1.12.13}$$

The magnetization of the medium can be neglected. The magnetic induction can be written as

$$\boldsymbol{B} = \mu_0 \boldsymbol{\mathcal{H}}. \tag{1.12.14}$$

The losses in the medium can be accounted for by introducing a fictitious conductivity so that

$$\boldsymbol{J} = \sigma \mathcal{E}. \tag{1.12.15}$$

The linear polarization is taken into account by considering the dielectric permittivity ϵ in the usual way, so that

$$\boldsymbol{D} = \epsilon \mathcal{E} + \boldsymbol{P}'. \tag{1.12.16}$$

From Eq. (1.12.11), (1.12.12), (1.12.15) and (1.12.16) we get

$$\nabla \times \nabla \times \mathcal{E} = -\mu_0 [\sigma \frac{\partial \mathcal{E}}{\partial t} + \epsilon \frac{\partial^2 \mathcal{E}}{\partial t^2} + \frac{\partial^2 \boldsymbol{P'}}{\partial t^2}]. \tag{1.12.17}$$

Since we have

$$\nabla \times \nabla \times \mathcal{E} = -\nabla^2 \mathcal{E} + \nabla (\nabla . \mathcal{E}) \cong -\nabla^2 \mathcal{E}.$$

We assume that $\nabla . \mathcal{E} = 0$, for small charge density in the medium, $\nabla . \boldsymbol{D} = 0$.

$$\nabla^2 \mathcal{E} - \frac{\sigma}{\epsilon c^2} \frac{\partial \mathcal{E}}{\partial t} - \frac{1}{c^2} \frac{\partial^2 \mathcal{E}}{\partial t^2} = \frac{1}{\epsilon c^2} \frac{\partial^2 \boldsymbol{P'}}{\partial t^2} \tag{1.12.18}$$

We have used, $\boldsymbol{c^2} = \frac{1}{\epsilon \mu_0}$. The linear part of the polarization is taken care of by the dielectric constant. The non-linear polarization on the right hand side is taken as a source in the wave equations. If we consider one dimensional propagation of the wave along the direction z, we can write down the equation for one component of the field as

$$\frac{\partial^2 \mathcal{E}}{\partial z^2} - \frac{\sigma}{\epsilon c^2} \frac{\partial \mathcal{E}}{\partial t} - \frac{1}{c^2} \frac{\partial^2 \mathcal{E}}{\partial t^2} = \frac{1}{\epsilon c^2} \frac{\partial^2 P'}{\partial t^2}. \tag{1.12.19}$$

We assume that the electric field at a frequency ω is given by the expression

$$\mathcal{E}(z,t) = \frac{1}{2} [\mathcal{E}_0(z) \exp\{i(\omega t - kz)\} + c.c.]. \tag{1.12.20}$$

The amplitude of the non-linear polarization can be written as

$$P'(z,t) = \frac{1}{2} [P'(z) \exp\{i(\omega t - kz)\} + c.c.]. \tag{1.12.21}$$

Substituting in the equation (1.12.18) and neglecting the second derivative term with respect to z we get

$$2 \frac{\partial \mathcal{E}_0}{\partial z} + \frac{\sigma}{\epsilon c^2} \mathcal{E}_0 = -i \frac{\omega}{\epsilon c} P'. \tag{1.12.22}$$

This equation is valid for a frequency ω and is the basic working equation for non-linear effects.

In the case of SHG, we have from Eq. (1.12.20) and (1.12.21) the total electric field for the fundamental and the second harmonic wave as

$$\mathcal{E}(z,t) = \frac{1}{2}[\mathcal{E}_0(z,\omega)\exp\{i(\omega t - kz)\}$$

$$+\mathcal{E}_{0SH}(z,2\omega)\exp\{i(2\omega t - k_2 z)\} + c.c.]. \qquad (1.12.23)$$

Similarly the polarization is also given as

$$P'(z,t) = \frac{1}{2}[P'(z,\omega)\exp\{i(\omega t - kz)\}$$

$$+P'(z,2\omega)\exp\{i(2\omega t - k_2 z)\} + c.c.]. \qquad (1.12.24)$$

By substituting these expressions into the working equation Eq. (1.12.22) we can get the non-linear polarization. It shows the presence of an additional frequency at twice the frequency of the incident wave.

1.12.5 Optical Parametric Oscillator

The discussions on second harmonic generation may be extended to the case of optical parametric oscillation. It may be noticed that in the case of mixing of two harmonic waves at frequency ω_1 and ω_2 a sum frequency is generated at a frequency $\omega_3 = \omega_1 + \omega_2$. In this case the conservation of energy and momentum demands that

$$\hbar\omega_1 + \hbar\omega_2 = \hbar\omega_3 \qquad (1.12.25)$$

$$\hbar\boldsymbol{k}_1 + \hbar\boldsymbol{k}_2 = \hbar\boldsymbol{k}_3 \qquad (1.12.26)$$

The condition on the wave vectors satisfy the phase matching condition described in the case of SHG.

The optical parametric oscillation is a process which is reverse of sum frequency generation. Instead of the generation of a new wave with a frequency as the sum of two different incident waves with two different frequencies, a wave at frequency ω_3 may split into two waves at two different frequencies ω_1 and ω_2 called the idler and signal waves. The frequency and the wave vectors of the generated waves also obey the conservation conditions expressed by Eq. (1.12.1) and (1.12.2).

The physical process may be de described as follows. We consider a strong wave at frequency ω_3 and a weak wave at frequency $\omega_1$to be present in a non-linear crystal. The strong wave is the pump frequency source provided by a laser. The weak radiation, called the signal, is not externally applied, but it may be present in the system as noise radiation. As a result of the non-linear interaction in the crystal the wave at frequency ω_3 will beat with the wave at frequency $\omega_1$to produce a polarization component at the beat frequency $\omega_2 = \omega_3 - \omega_1$. Thus the energy will be transferred from the pump wave to the signal and idler wave. The difference between SHG and OPO is that in the case of SHG, a strong pump wave is necessary to produce a weaker wave at double the frequency. In the case of OPO, another wave is involved. It need not be externally applied, as stated earlier. The wave at the frequency $\omega_1$will be amplified. The other wave at frequency ω_2 will also be amplified when the crystal is placed between the mirrors that are perfectly reflecting (Fig. 1.37). One can generate coherent beams from the noise in the system like that in a laser oscillator. The non-linear crystal should be pumped by an appropriately focused pump beam and placed within a resonator. The oscillation will start when the loss in the crystal and the resonator is compensated by the gain from parametric oscillation process. Hence a threshold power of the pump is required. The oscillation at both the frequencies will start following the conditions given by the Eq. (1.12.1) and

(1.12.2). In this case the phase matching conditions are followed. The mirrors can be highly reflecting at one frequency (singly resonant) or both the frequencies (doubly resonant).

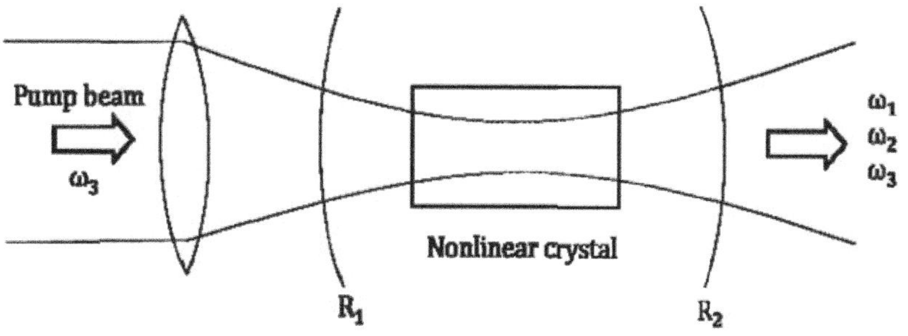

Fig. 1.37 Optical Parametric Oscillator exhibiting the generation of two new radiations.

The operation can be possible with both the cw and the transient sources of laser beams. For the singly resonant case higher pump power is necessary, but more stable output is obtained. The optical parametric oscillation producing coherent radiation beams has been obtained in the range of wavelength covering the visible to the near infrared region using beta barium borate (BBO),lithium niobate and other non-linear crystals.

Inside the crystal, all three radiations with three different frequencies are present. We can write their electric fields as

$$\mathcal{E}_3(z,t) = \frac{1}{2}[\mathcal{E}_{30}\ \exp\{i(\omega_3 t - k_3 z)\} + c.c], \tag{1.12.27}$$

$$\mathcal{E}_2(z,t) = \frac{1}{2}[\mathcal{E}_{20}\ \exp\{i(\omega_2 t - k_2 z)\} + c.c.], \tag{1.12.28}$$

$$\mathcal{E}_1(z,t) = \frac{1}{2}[\mathcal{E}_{10}\ \exp\{i(\omega_1 t - k_1 z)\} + c.c.]. \tag{1.12.29}$$

We have considered the complex electric field amplitudes and the complex conjugate terms are added. We consider non-linear effects in the crystal and hence we have to consider the real electric fields.

The non-linear polarization term can be written in this case as

$$P' = \epsilon_0\, \chi' \textstyle\sum_{i=1,3} \mathcal{E}_i^{\,2} \tag{1.12.30}$$

Substituting the expressions of the electric fields in the polarizations, we can collect the non-linear polarization terms oscillating at the three frequencies and they are given as follows

$$P'_3(z,t) = \epsilon_0\, \chi'[\mathcal{E}_1\, \mathcal{E}_2 \exp\{i(\omega_3 t - [k_1 + k_2]z)\} + c.c.] \tag{1.12.31}$$

$$P'_1(z,t) = \epsilon_0\, \chi'[\mathcal{E}_3\, \mathcal{E}_2 \exp\{i(\omega_1 t - [k_3 - k_2]z)\} + c.c.] \tag{1.12.32}$$

$$P'_2(z,t) = \epsilon_0\, \chi'[\mathcal{E}_1\, \mathcal{E}_3 \exp\{i(\omega_2 t - [k_3 - k_1]z)\} + c.c.] \tag{1.12.33}$$

These non-linear polarization terms are to be used in the working equation (Eq. 1.12.22)) for the non-linear effect for obtaining the resultant field.

The phase matching condition is given by Eq. (1.12.2). The phase mismatch is defined by

$$\Delta k = k_3 - k_1 - k_2 \tag{1.12.34}$$

For achieving efficient non-linear polarization, the phase mismatch should be zero. This mismatch arises because there is a difference in the speed of the non-linear polarization that is generating the non-linear effect and that of the electric field that is generating it. This is clear from Eq. (1.12.27-29) and (1.12.31-33). The speed for the electric field component at frequency ω_1 is ω_1/k_1. But the speed of the non-linear polarization at the same frequency is $\omega_1/(k_3 - k_2)$. For efficient generation of the non-linear wave the source polarization and the generating electric field must travel at the same speed.

The major advantage of a parametric oscillator is the possibility of achieving a tunable output laser. A laser beam in the visible region incident on the crystal can generate a wave with a tunable output in the infrared region. Hence this oscillator has been used as a tunable output source in many applications, such as spectroscopy. They offer wide range of tunability. For a particular pump frequency, the signal and idler waves are amplified. The frequencies of the waves that are amplified depend on the phase matching condition. This condition is determined by the refractive index of the crystal at the respective frequencies. Hence any method that can change the refractive index of the crystal can be used to tune the frequency of the signal and idler waves. Thus the frequencies may be tuned by changing the temperature of the crystal that can change the refractive index. They may also by tuned by applying an electric field which changes the refractive indices by means of electro-optic effect. Tuning is also possible by changing the orientation of the crystal, if one of the waves is an extraordinary wave.

Problems

1.1. In a Young's double slit experiment show that the maximum distance l of a point from the axis for disappearance of interference fringes should satisfy the condition

$$l \approx \frac{\lambda D}{2d}$$

where D is the distance of the source from the double slit and d is the distance between the slits (Fig. 1. 7).

1.2. Show that the ratio of the number of spontaneous to stimulated emission is $(n_1 - n_2)/\, n_2$, where n_1 and n_2 are the populations of the lower and upper levels as defined in Eq. (1.2.4).

1.3. In an optical source, the temperature T is nearly 1000° K and the frequency ω is around 10^6 GHz. Using these data explain why the emission from usual optical light sources is incoherent.

1.4. Calculate the acceptor concentration for minimum response time of 2 ns, if the probability of recombination B=$10^{-9}\frac{cm^3}{sec}$. What can be done to have faster response?

1.5. If the input current i to the diode is modulated sinusoidally at a frequency ω with an amplitude i_0 and the electron density oscillates at the same frequency and Γ is the decay probability as defined in the text show that the electron density can be written as

$$N(t) = \frac{i_0}{e(\Gamma^2+\omega^2)^{1/2}} e^{i(\omega t-\theta)} \qquad \text{where, } \tan\theta = \frac{\omega}{\Gamma}.$$

1.6. Consider a three dimensional cavity as a cube of volume $V = L^3$. Show that the number of modes with frequencies upto the value $v = ck/2\pi$ is $\frac{8\pi v^3 V}{3c^3}$. [$k_i = m_i\left(\frac{L}{\pi}\right), i = x, y, z$, the surface in k-space corresponding to the maximum k-value is assumed to be a sphere of radius k and there are two possible polarizations for each spatial mode].

1.7. Calculate the spectral mode density defined as the number of modes per unit frequency interval per unit volume in the three dimensional cavity as defined in the previous problem. Also obtain the number of modes for a small frequency interval dv.

1.8. A semiconductor laser emits at a wavelength of 820 nm and has a longitudinal coherence length of 1 mm. Find the frequency and wavelength width.

1.9. An Argon ion laser has a cavity length of 1 meter and emits at 514 nm. Find the mode spacing, mode width, Quality factor and mode finesse of the laser if the reflectivities are $\rho_2 = 0.9$ and$\rho_1 = 0.99$ and assume the refractive index to be 1.0.

1.10. If the laser in the Problem 1.9 oscillates in a single mode, find the coherence length L_c.

1.11. Substituting the trial solution in the form of Eq. (1.9.20) into Eq. (1.9.19) show that the wave amplitude $\psi(r, z)$ can be written in the form of Eq. (1.9.21).

1.12. Show that the laser spot size at the mirrors can be expressed by

$$w^2\left(\frac{L}{2}\right) = \left(\frac{\lambda r}{\pi}\right)/\left(\frac{2r}{L} - 1\right)^{1/2}.$$

1.13. What happens to the spot size at the mirror and the waist size when $r \to \frac{L}{2}$? Where is the centre of curvature of the two mirrors in this case?

1.14. Justify the fact that a cavity with perfectly flat mirrors will be lossy.

1.15. In a He Ne laser with radii of curvature much larger than the cavity length, if the resonator length is reduced from 100 cm to 10 cm, how will the spot size change at the centre?

1.16. Write down the rate equations for population of a two-level atom interacting with a resonant radiation field. Consider absorption from the lower level and both spontaneous and stimulated emission from the upper level. Obtain the ratio of populations in the steady state and show that it is not possible to have population inversion in a two level atom in the steady state.

1.17. Show that in the microwave frequency region 7-40 Ghz, the number of spontaneous emissions is negligible at the operating temperature.

1.18. What is the threshold pumping rate when the rate of spontaneous transition from level 2 to level 1 is small? What is the population difference $N_2 - N_1$ under this condition?

1.19. In a Ruby laser the spontaneous emission life time is 3 μsec, the atom population density is 1.9 x 10^{19} per c.c. and the average population inversion density is 1.5 x 10^{17} per c.c below laser threshold., calculate the pumping rate per sec from the lower laser level.

1.20. In a three-level atom the life time of the atoms in level 2 is 1 μsec. The decay of atoms from level 3 to the level 2 is very fast. If the atom density is 10^{19} $per\ c.c.$ calculate the threshold value of the pumping rate from level 1 per sec per unit volume.

1.21. In a doubly resonant oscillator the pump (ω_p), signal (ω_1) and idler (ω_2) frequencies have to satisfy the condition

$$\omega_p n_p = \omega_1 n_1 + \omega_2 n_2$$

If the pump and signal wavelengths are 1.2 μm, and 1.5 μm, find the idler wavelength, when the refractive indices of the pump, signal and idler waves are 1.5, 1.6 and 1.7 respectively.

Further reading

1. K. Thyagarajan and A. Ghatak, Lasers, Fundamentals and Applications, Springer, Berlin, 2010.
2. R. S. Quimby, Photonics and Laser Applications, Wiley-Interscience, New York, 2006.
3. O. Svelto, Principles of Lasers, Springer, Berlin, 2010.
4. A.Yariv, Optical Electronics and Modern Communications, 5th edition, Oxford University Press, 1977.
5. C. H. Townes, Nobel Lectures in Physics, December 11, 1964.
6. A. M. Prochorov, Nobel Lectures in Physics, December 11, 1964.
7. N. G. Basov, Nobel Lectures in Physics, December 11, 1964.
8. A. Yariv, Quantum Electronics, 2nd Edition, John Wiley,1975.
9. K. Shimoda, Introduction to Laser Physics, Springer, Berlin, 1986.
10. W T Silfvast, Laser Fundamentals, Second Edition, Cambridge University Press, Cambridge, 2004.

Chapter 2
Semi-classical Theory of Atom-Field Interaction

2.1 Semi-classical treatment

Quantum mechanics describes an atom as a quantized system with a set of discrete energy levels. Each energy level is designated by a set of quantum numbers like the principal quantum of the Rydberg level and the quantum numbers representing the orbital, spin and total angular momentum. Depending on the existence of nuclear spin, there may be further assignments of levels because of the hyperfine splitting. An atom in the presence of an external field may have splitting of the energy levels and this may lead to further new designations by quantum numbers. A molecule will have a different set of quantum numbers arising from its rotational and vibrational motions. Atoms in a solid have energy levels in the form of a band. In all the cases, when an atom or an atomic system interacts with an electromagnetic radiation with a frequency ν, the radiation will interact with the atom such that the energy difference between the two levels equals the energy of the incident photon. This is the condition of resonance. All other energy levels remain unaffected. Thus, although an atom or an atomic system has many energy levels, for the interacting radiation with a particular frequency *the atom can be considered as a two level system.*

In our treatment in this chapter we shall consider the atom as a quantum mechanical system. But the electromagnetic radiation will be treated as classical waves described by the Maxwell equations. For this purpose, the treatment is called semi-classical. Electromagnetic radiation can also be quantized and the interaction can be treated quantum mechanically. We shall consider the quantum theory of radiation and the atom field interaction in Chapters 6 and 7.

2.2 Atom Field Interaction

We consider a two level atom having energy $E_n = \hbar\omega_n, (n = a, b)$ described by the eigenvalue equations

$$H_0\varphi_n = E_n\varphi_n, \tag{2.2.1}$$

where φ_n are the respective time-independent eigen functions and H_0 is the Hamiltonian of the two level atoms (Fig. 2.1). In terms of the resonance frequency $\omega = \omega_b - \omega_a$, the Hamiltonian can be written as

$$H_0 = \hbar/2 \begin{pmatrix} \omega & 0 \\ 0 & -\omega \end{pmatrix}. \tag{2.2.2}$$

The zero value of energy is set at the mean value of the two energies. The atom can be described by a general wave function expressed as the linear superposition of the two eigenfunctions φ_a and φ_b as

$$\psi(r,t) = \sum_n C_n(t)\ \varphi_n(\mathrm{r})\, e^{-i\omega_n t}. \tag{2.2.3}$$

In the presence of an electromagnetic field the Hamiltonian of the atom field system can be written as

$$H = H_0 + V, \tag{2.2.4}$$

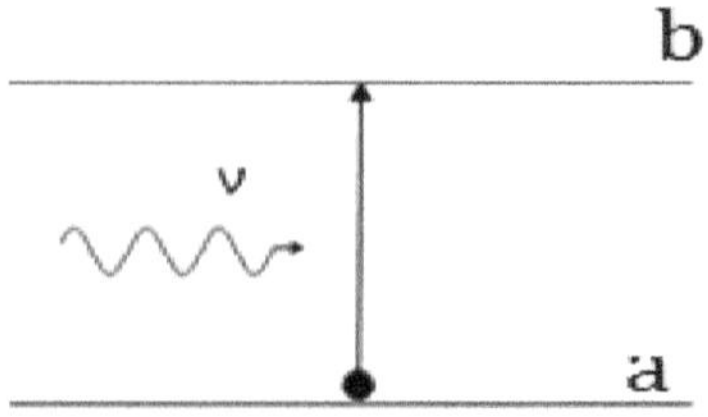

Fig. 2.1 Energy levels of a two-level atom with energies $E_b = \hbar\omega_b$ and $E_a = \hbar\omega_a$ interacting with electromagnetic radiation of frequency ν.

where V is the atom-field interaction operator that depends both on the position and time coordinates.

$$V = -\mu\, \mathcal{E}_0 \cos \nu\, t, \tag{2.2.5}$$

μ is the transition dipole operator of the atom, $\mathcal{E}_0$ is the electric field amplitude of the applied field and ν is the field frequency. We have considered the component of the electric dipole moment in the direction of the electric field. In the presence of the applied field the atomic wave function will evolve in time and the coefficients C_n will be time dependent functions. We have to consider the time dependent Schroedinger equation as

$$i\hbar \frac{\partial\psi}{\partial t} = H_0\psi. \tag{2.2.6}$$

Substituting Eq. (2.2.3) into Eq. (2.2.6) and using Eq. (2.2.1) and (2.2.4) we get

$$i\hbar \sum_n (\dot{C}_n - i\omega_n C_n)\, e^{-i\omega_n t}\varphi_n(r) = \sum_n C_n(t)(\hbar\omega_n + V)e^{-i\omega_n t}\varphi_n(r). \tag{2.2.7}$$

For the two-level atom $n = a, b$. Multiplying Eq. (2.2.7) by φ_b^* and φ_a^* we obtain the two equations respectively,

$$\dot{C}_a = -\frac{i}{\hbar} C_b V_{ab} e^{-i(\omega_b - \omega_a)t}, \tag{2.2.8}$$

$$\dot{C}_b = -\frac{i}{\hbar} C_a V_{ba} e^{-i(\omega_a - \omega_b)t}. \tag{2.2.9}$$

We have used the eigenvalue equation (2.2.1) and the interaction matrix element

$$V_{nk} = \int \varphi_n^*\, V\, \varphi_k dv. \tag{2.2.10}$$

We consider the electric dipole interaction described by Eq. (2.2.5) so that

$$V_{nk} = -\int \varphi_n^*\, \mu\varepsilon_0\, \cos \nu t\, \varphi_k dv. \qquad 2.2.11)$$

We consider $\mu_{kn} = \int \varphi_n^*\, \mu\, \varphi_k dv$ as the dipole operator matrix element between the states k and n. For the two level atom, $\mu_{ab} = \mu_{ba} = \mu$(say), since the non-diagonal matrix elements are equal. It should be noted here that μ is now a matrix element, so it is a number, but $\boldsymbol{\mu}$ mentioned earlier was an operator. The diagonal matrix eleme*nts are zero* since the dipole operator has odd parity and the atomic wave functions φ have a definite parity. Hence the electric dipole interaction for the two-level atom is

$$V_{ab} = -(\mu\, \varepsilon_0/2)(e^{i\nu t} + e^{-i\nu t}).$$

Equations (2.2.8) and (2.2.9) may be written as

$$\dot{C}_a = (i\mu\varepsilon_0/2\hbar)\,[\,e^{-i(\omega-\nu)t} + e^{-i(\omega+\nu)t}]C_b, \quad (2.2.12)$$

$$\dot{C}_b = (i\mu\varepsilon_0/2\,\hbar)\,[\,e^{i(\omega-\nu)t} + e^{i(\omega+\nu)t}]\,C_a. \quad (2.2.13)$$

These equations are known as the rate equations for the C coefficients. They can be solved by approximation methods for the weak field case and by using exact Rabi method for strong field cases.

2.3 Weak Field Case

If the field is weak, the atom field interaction energy is small and hence the effect of V may be considered as a perturbation. As mentioned in section 2.1, in the visible or higher frequency region the number of atoms in the upper state is much small compared to that in the lower state. Hence the number of atoms transferred to the upper state from the lower state by transition caused by a weak field will be extremely small compared with the total number of atoms available in the ground state. This means that the total number of atoms in the lower state will remain effectively unchanged, although a significant number may undergo transition so that they can be experimentally observed. We consider a case where all the atoms are in the lower state so that

$$C_a\,(0) = 1 \text{ and } \; C_b(0) = 0. \quad (2.3.1)$$

Since the change caused by the field is small, we assume that C_a (t) will remain the same for all time as at the initial time t=0, but there will be change in C_b (t). Thus,

$$\dot{C}_a = 0, \quad (2.3.2)$$

$$\dot{C}_b = (\varepsilon_0/2\hbar)\,[\,e^{i(\omega-\nu)t} + e^{i(\omega+\nu)t}]. \quad (2.3.3)$$

This assumption may appear somewhat strange, since C_a remains unperturbed, then why C_b will change. We can consider the numerical example. What actually happens is that $|C_a(t)|^2$ represents the population density. Usually in an atomic system at room temperature and atmospheric pressure, the population density is of the order of 10^{19} atoms per c.c. in the ground level and the field causes a small change in the upper level population that may increase from zero to a number 10^{15} (as an example). Hence the lower level population will be 10^{19} - 10^{15} = $10^{19}(1\text{-}10^{-4}) \sim 10^{19}$. It may be noted that $|C_a(t)|^2$ undergoes a negligible change, but the upper level population changes significantly for observation. *This is the basic justification of time dependent perturbation calculations of quantum mechanics. However, this assumption is not valid for radiation frequency in the microwave region.* For low resonance frequency, the population difference between the two interacting energy levels is small. Even in the presence of a weak radiation field saturation (when the populations of the two levels are almost the same) is attained easily.

After integrating eq. (2.3.3), we get

$$C_b(\text{t}) = (i\mu\varepsilon_0/2\hbar)\;[\frac{e^{i(\omega-\nu)t}}{i(\omega-\nu)} + \frac{e^{i(\omega+\nu)t}}{i(\omega+\nu)}]. \quad (2.3.4)$$

In the case of absorption, the lower level has more population. Since $\omega_b > \omega_a$, ω is positive. In the case of resonance, $\omega \cong \nu$, hence $\omega - \nu$ is very small and $\omega + \nu$ is nearly 2ω, so the first term dominates and the second term can be neglected. This is equivalent to omitting one of the two terms of $\cos \nu t$ in the electric field. This is known as **Rotating Wave Approximation (RWA).** It describes a process in which the atomic resonance frequency ω closely follows the electromagnetic field frequency ν. If the harmonic frequencies ω and ν can be described by the circular rotation of a corresponding vector then ν will closely follow ω or the two vectors will rotate together with the same phase difference. It will be shown later that this approximation can be explained also by the principle of conservation of energy. Eq. (2.3.4) leads to

$$|C_b\,(t)|^2 \quad = (\mu\,\mathcal{E}_0/2\hbar)^2\,\frac{\sin^2(\omega-\nu)t/2}{[(\omega-\nu)/2]^2}\,. \tag{2.3.5}$$

This equation represents the probability of absorption of the two-level atom. It has a maximum value at resonance. The probability of absorption is plotted as function of detuning in Fig. 2.2. In addition to the central maximum there are other smaller maxima at a particular time t. In the case of emission, we have more population in the upper state, so we should consider the initial condition as $C_a(0)$ =0 and $C_b(0) = 1$. In this case, we can calculate $|C_a\,(t)|^2$ that represents the probability of emission from

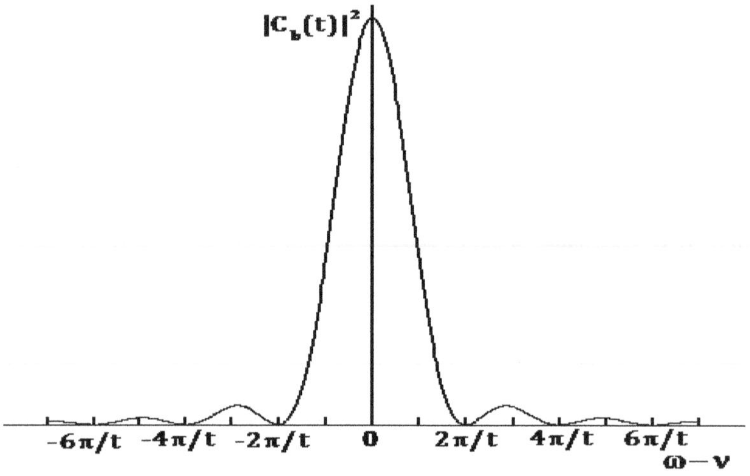

Fig. 2.2 Absorption probability as a function of detuning.

the upper state to the lower state. These probability terms represent the Einstein coefficient B of emission or absorption. It is found that the probability of emission is equal to the probability of absorption.

In the above, we have discussed how the probabilities of stimulated emission and stimulated absorption can be described by semi classical treatment of atom-field interaction. *We cannot obtain the spontaneous emission using the same model.* We shall show later how this can be derived from quantum theory of radiation. Einstein used thermodynamic equilibration model to obtain a relation between the A and B coefficients as discussed in Section 1.2.

2.4 Broadening of spectral lines

The observed spectrum due to spontaneous emission from a collection of atoms is not strictly monochromatic, but it is spread over a certain range of frequencies. Similarly, the absorption spectrum also shows that the atom does not absorb at a single frequency, but it is spread over a band of frequencies around the resonance frequency. This implies that the energy is spread over a band of energy. It may be mentioned here that the energy levels have natural broadening arising from Heisenberg uncertainty relation.

The interaction strength at different frequencies is a function of frequency and this function may be called a line shape function $\xi(\omega)$. A simple explanation of the line shape is based on the finite life time of the atom in the state. This gives rise to a broadening of the quantum mechanical energy. This is one reason of broadening. The observed spectrum also shows broadening originating from other mechanisms. We shall discus some of them. The line shape function is normalized to unity, so that

$$\int \xi(\omega)d\omega = 1. \tag{2.4.1}$$

Thus $\xi(\omega)$ has the unit of sec. The total number of stimulated emissions per unit time per unit volume from level b to level a (Eq. (1.2.3)) is

$$\mathbb{N}_{ba} = \int B_{ba}U(\omega)\, N_b\xi(\omega)d\omega. \tag{2.4.2}$$

This is because among the N_b atoms per unit volume, only $N_b\xi(\omega)d\omega$ atoms per unit volume can interact with the radiation in a frequency range ω and $\omega + d\omega$. If the radiation extends over a broad range of frequencies, $U(\omega)$ is much broader than $\xi(\omega)$. For a monochromatic radiation, $U(\omega)$ is much narrower compared to $\xi(\omega)$. The operation characteristics of a laser, such as the threshold inversion population and mode behavior, discussed in the Chapter 1, depend very much on the line broadening.

Broadening may be divided into two types, called homogeneous or inhomogeneous broadening. In certain cases like natural or collision broadening, all atoms in an ensemble respond to the interacting radiation in an identical manner. The transition probability of radiation of a certain frequency for each atom is the same and the atoms cannot be distinguished from each other. Hence, the broadening is homogeneous. In the case of Doppler broadening, the atoms see a shifted radiation frequency depending on their velocities. Hence, the response of the atoms having different velocities can be different. This kind of broadening is called inhomogeneous broadening. In the case of solids, local defects or inhomogeneities can cause shift of the frequency leading to inhomogeneous broadening.

2.4.1 Natural broadening

From Eqs. (1.2.10-14), we find that the change of energy due to spontaneous emission from the upper state per unit volume per unit time or the intensity of spontaneous radiation is given by

$$I(t) = A_{ba}N_b(0)\hbar\omega e^{-t/\tau}. \tag{2.4.3}$$

Since the intensity is proportional to the square of the electric field, the electric field associated with the spontaneous emission is defined by

$$\mathcal{E}(t) = \mathcal{E}(0)e^{i\omega t}e^{-t/2\tau}, \tag{2.4.4}$$

$\mathcal{E}(0)$ is the electric field amplitude at t=0. The above shows that the oscillatory electric field decreases exponentially with time. In order to calculate the power spectrum we have to find the Fourier transform of the time dependent function. So the electric field is

$$\mathcal{E}(\nu) = \mathcal{E}(0)\int_0^\infty e^{i\omega t}e^{-t/2\tau}e^{-i\nu t}dt = \mathcal{E}(0)\frac{1}{i(\nu-\omega)+1/2\tau}. \tag{2.4.5}$$

The power spectrum is proportional to $|\mathcal{E}(\nu)|^2$. Hence we get

$$\xi(\nu) = C\,\frac{1}{(\nu-\omega)^2+1/4\tau^2}.$$

C is a constant of proportionality and is determined by the normalization condition of $\xi(\nu)$ (Eq. 2.4.1). This leads to

$$\xi(\nu) = \frac{1}{2\pi\tau} \frac{1}{(\nu-\omega)^2 + 1/4\tau^2} \tag{2.4.6}$$

Thus the line shape arising from natural broadening is Lorentzian in nature. Hence the Half Width at Half Maximum (HWHM) is $1/2\tau$ and the full width FWHM, $\Delta\nu_N = 1/\tau$ (Fig. 2.3). If the spontaneous emission life time is 10 ns for an atomic transition, the line width FWHM is 100 MHz. If the frequency is expressed in the unit of radian/second this value is $2\pi \times 100\ MHz$. In the case of hyperfine components of D type transitions of Rb, the natural line width can be of the order of 5 MHz.

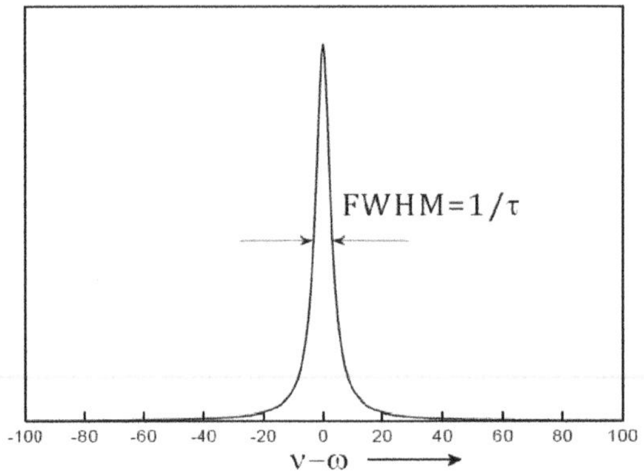

Fig. 2.3 Plot of Lorentzian line shape function $\xi(\nu)$ against frequency in the case of natural broadening with the peak centered at the atomic resonance frequency ω.

2.4.2 Collision Broadening

A single stationary atom emits a monochromatic beam of radiation having a fixed wavelength. If another atom comes close to it, its energy levels are perturbed by the interaction with the second atom. For this reason, it will emit a monochromatic radiation having a shifted frequency or wavelength. This occurs when the two atoms are close and collide with each other. The emitted wavelength reverts back to the original value after the second atom moves away. After this event, another atom may also come and collide, so that the wavelength changes again. If the duration of each collision is small compared to the time between successive collisions, we can assume that the emitted wave from the atom consists of a series of photons having a distribution of wavelengths. The time between collisions, T depends on the mean free path and is of the order of $10^{-6}\ sec$ for atoms at room temperature. Whereas the duration of collision T_c is determined by the distance between the atoms at the moment of collision and is of the order of $10^{-13}\ sec$. Hence the collision is almost instantaneous. After each collision the phase of the atom changes and this is a random process. So the emitted wave is not monochromatic. This is known as collision broadening.

If $P(T)dT$ is the probability that the atom suffers a collision in the time interval between T and T + dT, then the mean time between two collisions is

$$T_0 = \int_0^\infty T P(T) dT \tag{2.4.7}$$

and

$$\int_0^\infty P(\mathrm{T})d\mathrm{T} = 1. \qquad (2.4.8)$$

The collision probability varies exponentially with time and can be expressed as the normalized function

$$P(\mathrm{T}) = \frac{1}{\mathrm{T}_0} e^{-\mathrm{T}/\mathrm{T}_0}. \qquad (2.4.9)$$

The power spectrum as a function of atomic transition frequency for the time interval 0 to T is given by Eq. (2.3.5) as

$$I(\nu, \mathrm{T}) = (\mu \mathcal{E}_0/\hbar)^2 \frac{\sin^2(\nu-\omega)\mathrm{T}/2}{(\nu-\omega)^2}. \qquad (2.4.10)$$

The line shape function $\xi(\nu)$ is then

$$\xi(\nu) = \int_0^\infty I(\nu, \mathrm{T})\, P(\mathrm{T})d\mathrm{T}. \qquad (2.4.11)$$

This leads to

$$\xi(\nu) = (\mu\, \varepsilon_0/\hbar)^2\, \frac{1}{2}\, \frac{1}{(\nu-\omega)^2 + 1/\mathrm{T}_0{}^2}. \qquad (2.4.12)$$

The FWHM for this Lorentzian function is

$$\Delta\nu = 2/\mathrm{T}_0\,. \qquad (2.4.13)$$

Hence the width of the Lorentzian functions of natural and collision broadening are additive. So the line width of the **Lorentzian** (Fig. 2.4) function including natural and collision broadening is given by

$$\Delta\nu_L = \frac{2}{\mathrm{T}_0} + 1/\tau \qquad (2.4.14)$$

The value of collision broadening depends on the atom density. In the case of gaseous system it depends on the gas pressure. Buffer gases like inert nitrogen atoms may also be added in order to reduce the line broadening due to collision.

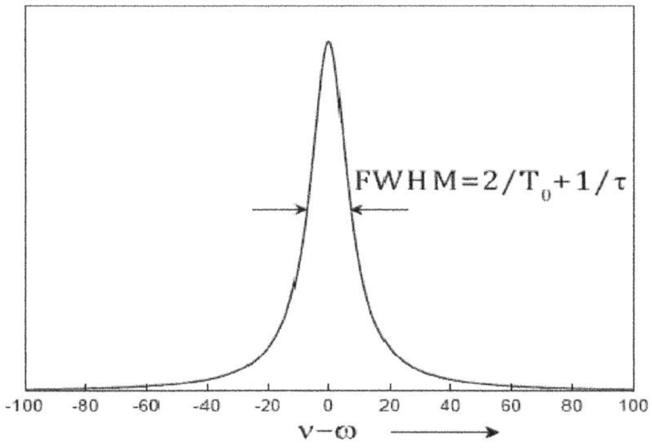

Fig.2.4 Plot of Lorentzian line shape function by including both natural and collision broadening and the peak centered at the atomic resonance frequency.

2.4.3 Doppler broadening

In the gas phase, the atoms move in random directions with different velocities. An atom, moving with a particular velocity v, will see the radiation, coming along the same axis from opposite direction, with a Doppler-shifted frequency

$$\nu_D = \nu\left(1 - \frac{v}{c}\right). \tag{2.4.15}$$

For resonance we need that the atomic transition frequency is equal to ν_D or $\omega = \nu_D$. Hence

$$\nu = \omega/(1 - \frac{v}{c}) \cong \omega(1 + \frac{v}{c}). \tag{2.4.16}$$

Because we can assume $v \ll c$. The probability that the atom can have velocity lying between v and $v + dv$ is given by the Maxwell-Boltzmann distribution as

$$P(v)dv = (\frac{M}{2\pi k_B T})^{1/2} \exp\left(-\frac{Mv^2}{2k_B T}\right) dv. \tag{2.4.17}$$

Hence the probability that the radiation frequency lies between ν and $\nu + d\nu$ ($dv = \frac{c}{\omega} d\nu$)is

$$\xi(\nu)d\nu = \frac{c}{\omega} (\frac{M}{2\pi k_B T})^{1/2} \exp\left(-\frac{Mc^2(\nu-\omega)^2}{2\omega^2 k_B T}\right) d\nu, \tag{2.4.18}$$

where, we have used the expression of velocity from Eq. (2.4.15). Thus the Doppler broadened line shape function (Eq. 2.4.18) is given by a **Gaussian** (Fig. 2.5) function. It has a peak at frequency ω and the FWHM is

$$\Delta\nu_D = 2\omega(\frac{2k_B T}{Mc^2} ln2)^{1/2} . \tag{2.4.19}$$

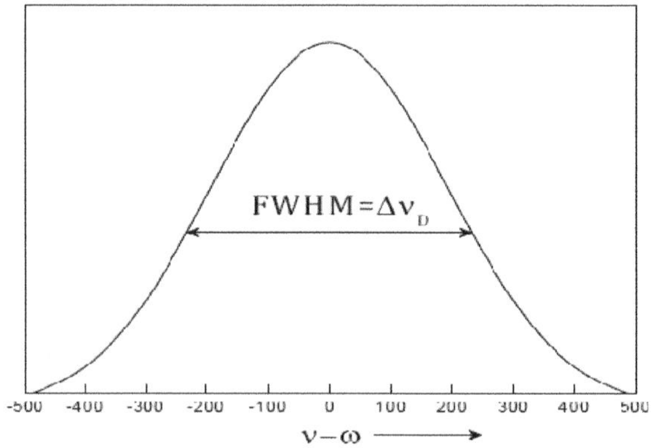

Fig. 2.5 Plot of Gaussian Line shape function in the case of Doppler broadening with the peak centered at the atomic transition frequency.

At normal temperature (T= 300 K), the Doppler broadened FWHM for Rubidium D-type transition can be nearly 450 MHz. In the case of sodium discharge, the Doppler width of sodium D_1 line at 589.0 Å is 1.7 GHz or a wave length width of 0.02 Å . For the 10.6 μm vibrational infrared transition of carbon dioxide, the Doppler width is 0.19 Å. The Doppler width is determined by the radiation frequency and the temperature and mass of the atom or molecule.

Voigt Profile

In any experiment with gaseous atoms, both collision and Doppler broadening will dominate, although there are other broadening mechanisms as well. The observed line shape will be a *convolution of a Lorentzian and a Gaussian profile*. This is known as a Voigt profile (Fig. 2.6).

The Lorentzian Line Shape function can be written by including both natural (Eq. (2.4.6)) and collision broadening (Eq. 2,4.12)) as

$$\xi_L(\nu) = (\mu\, \varepsilon_0/\hbar)^2\, \frac{1}{2}\, \frac{1}{(\nu-\omega)^2+(\frac{1}{2\tau}+1/T_0)^2}. \quad (2.4.20)$$

The Voigt profile can be expressed as

$$\xi_V(\nu) = \int \xi_L(\nu)\, \xi_D(\nu-\nu')d\nu'. \quad (2.4.21)$$

The function $\xi_D(\nu-\nu')$ is the Doppler line shape function as given by Eq. (2.4.18). In the usual experimental conditions Doppler broadening is much larger than the Lorentzian line width. So the Lorentzian line width gets masked by the Doppler width within the Voigt profile. It will be shown in the Chapter 8 that using non-linear spectroscopy we can extract the Lorentzian line shape that lies hidden in the Voigt profile which looks much closer to the Doppler line shape.

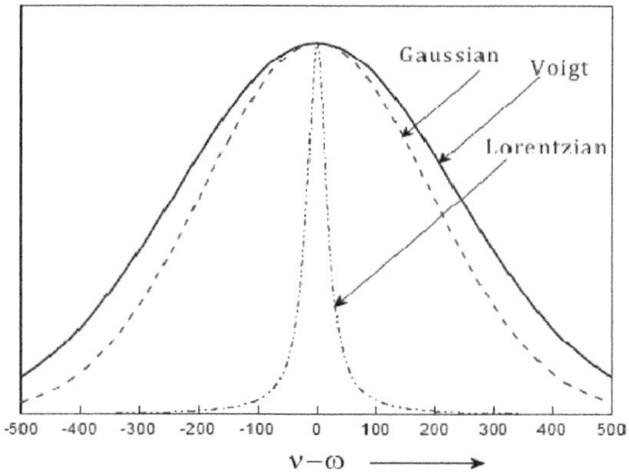

Fig. 2.6 Plot of Voigt line shape function that includes both the homogeneous broadening given by the Lorentzian function and the inhomogeneous broadening given by the Gaussian function.

2.5 Strong Field Case: Rabi Oscillation

In section 2.1, we discussed the interaction of a quantized atomic system with weak electromagnetic field that causes transition from one level to the other. In this case, change in the population of levels is small. In presence of the fields, both the stimulated absorption and emission processes are active. But usually, the absorption will dominate, because there are more particles in the lower state than in the upper state. If the field is strong, a much larger number of particles are transferred to the upper state. However, since the particle number in the upper state cannot exceed that of the lower state, the maximum number that can be transferred is such that there is saturation, so that $N_1 = N_2$.

In the case of strong field, the electric field amplitude ε_0 is large. Hence, the interaction energy is not small. In such a case, the interaction cannot be treated as perturbation. We shall follow Rabi's method of exact solution and use RWA.

We shall use trial solution for $C_a(t)$ in the form

$$C_a(t) = A\, e^{ift}. \tag{2.5.1}$$

Under RWA we can omit the terms with $\omega + v$, hence equations 2.2.12 and 2.2.13 can be written as

$$\dot{C}_a(t) = i\frac{\Omega}{2}e^{-i(\omega-v)t}C_b(t), \tag{2.5.2}$$

$$\dot{C}_b(t) = i\frac{\Omega}{2}e^{i(\omega-v)t}C_a(t). \tag{2.5.3}$$

We have defined the **Rabi frequency** as

$$\Omega = \frac{\mu \mathcal{E}_0}{\hbar}. \tag{2.5.4}$$

This term has the dimension of frequency and depends on the strength of the applied field and the transition dipole moment only. *It is, in no way, related to the oscillation frequency of the field or the resonance frequency of the atom.* Substituting the trial solution in Eq. (2.5.3) we get

$$\dot{C}_b = A\frac{i\Omega}{2}\, e^{i(\omega-v+f)t}. \tag{2.5.5}$$

Substituting the trial solution in Eq. (2.5.2) and taking time derivative we get

$$\dot{C}_b = A\frac{2f}{\Omega}\, i(\omega - v + f)e^{i(\omega-v+f)t}. \tag{2.5.6}$$

By comparing the last two equations, we can write

$$f^2 + (\omega - v)f - \frac{\Omega^2}{4} = 0. \tag{2.5.7}$$

So we get

$$f_{1,2} = -\frac{\omega-v}{2} \pm \sqrt{\frac{(\omega-v)^2+\Omega^2}{4}}. \tag{2.5.8}$$

Thus there are two solutions for f : f_1 and f_2. Hence the general solution for C_a can be written as

$$C_a(t) = a_1e^{if_1t} + a_2e^{if_2t}. \tag{2.5.9}$$

Substituting Eq. (2.5.9) into equation for $\dot{C}_a$ we get

$$C_b(t) = \frac{2}{\Omega}[\, a_1f_1e^{if_1t} + a_2f_2e^{if_2t}\,]e^{i(\omega-v)t}. \tag{2.5.10}$$

If we choose the initial condition such that all the particles are in the state a, then, as in the weak field case, we have from Eq. (2.3.1),

$$a_1 + a_2 = 1,$$

$$a_1 f_1 + a_2 f_2 = 0.$$

Hence

$$a_1 = -\frac{f_2}{f_1 - f_2}, \quad a_2 = \frac{f_1}{f_1 - f_2}. \tag{2.5.11}$$

From Eq. (2.5.8), it follows that the difference frequency is given by

$$F = f_1 - f_2 = [\,(\omega - v)^2 + \Omega^2\,]^{1/2} \tag{2.5.12}$$

and

$$f_1 f_2 = -\frac{\Omega^2}{4}. \tag{2.5.13}$$

From Eq. (2.5.10 - 2.5.12), we obtain

$$C_b(t) = i\frac{\Omega}{F} e^{i\frac{(\omega - v)t}{2}} \sin\frac{Ft}{2}, \tag{2.5.14}$$

$$|C_b\,(t)|^2 = \Omega^2 \frac{\sin^2[(\omega - v)^2 + \Omega^2]^{1/2} t/2}{[(\omega - v)^2 + \Omega^2]}. \tag{2.5.15}$$

A comparison of Eq. (2.5.15) with Eq. (2.3.5) for the weak field case shows that in case of small Ω, the strong field solution reduces to the weak field case. The absorption probability is similar to the curve shown in Fig. 2.2, except that the positions of the other maxima depend on the Rabi frequency in addition to the time. There is an additional term in the frequency in Eq. (2.5.15); this is represented by the power dependent Rabi frequency. In the case of high power, as may be available from a laser, the system shows a power dependent oscillation. It will be shown later that the additional term in F leads to **power broadening.** The same value for the emission probability may also be obtained for a state that is initially excited. This formula, called the **Rabi flopping** formula, was derived by I. I. Rabi for a spin ½ system. The probability that a spin-1/2 atom incident on a Stern-Gerlach apparatus could be flipped from the $\binom{1}{0}$ state to the $\binom{0}{1}$ state or from the $\binom{0}{1}$ state to the $\binom{1}{0}$ state in the presence of an applied radiofrequency magnetic field could be expressed by this formula. Rabi used this expression for a special apparatus to get a more accurate measurement of the magnetic moment of the atoms compared to what is possible with the original Stern-Gerlach apparatus.

We can show that at resonance ($\omega = v$)

$$|C_b\,(t)|^2 = \sin^2 \Omega t/2\,. \tag{2.5.16}$$

Hence $|C_b\,(t)|^2$ oscillates between 0 and 1 (Fig. 2.7). Since

$$|C_a\,(t)|^2 + |C_b\,(t)|^2 = 1 \tag{2.5.17}$$

the atom oscillates between the levels a and b. If at the initial time $t = 0$, the atom is in the lower level, we have

$$C_a\,(0) = 1 \text{ and } C_b(0) = 0.$$

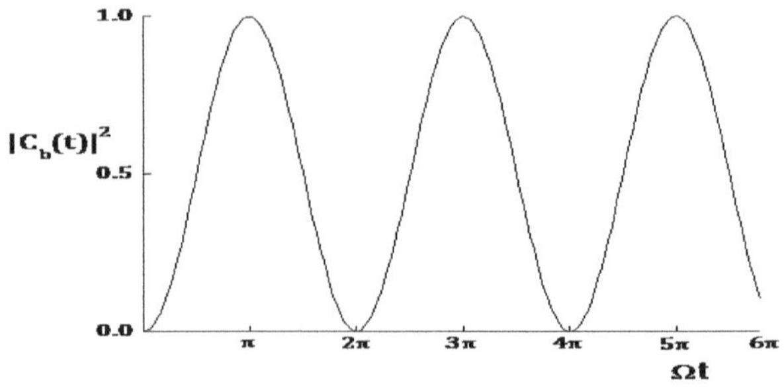

Fig. 2.7 Probability of absorption at zero detuning for strong field as a function of Ωt.

At a time $t_{1/2} = \frac{\pi}{\Omega}$,

$$C_a\,(t_{1/2}) = 0 \quad \text{and} \quad C_b(t_{1/2}) = 1.$$

Hence, at time $\frac{\pi}{\Omega}$ the atom switches to the level b, so we can say that at this time the system is inverted. The full period of oscillation may be defined as

$$T = \frac{2\pi}{\Omega} = \frac{2\pi\hbar}{\mu\mathcal{E}_0}. \tag{2.5.18}$$

Thus, the atom oscillates between the levels a and b with the time period T and frequency Ω. This phenomenon is called Rabi flopping. A pulse of duration $\frac{\pi}{\Omega}$ is known as the **π pulse**. This *pi*-pulse will take the atom from the lower state to the upper sate and the system will not revert back, if there is no spontaneous emission. This pulse changes the phase of the system by π. In presence of a continuous wave laser the system will flip back and forth with Rabi frequency.

Problems

2.1 A weak electromagnetic field is interacting with a two level atom. At initial time $t = 0$, all the atoms are in the excited state. Solve the rate equations with an appropriate initial condition and obtain the probability of stimulated emission to first order under RWA. Is this probability the same as the probability of absorption as given by Eq. (2.3.5)?

2.2 In the problem 2.1 indicate how you can obtain the solution to the next higher order.

2.3 If the radiation frequency is tuned on resonance ($\omega - \nu = 0$), show that the rate equations (2.2.12) and (2.2.13) can be solved exactly under RWA. Obtain the solutions.

2.4 A two-level atom with the energy levels described by Fig. 2.1 has an upper state decay constant γ (exponential decay). Write down the rate equations in presence of a weak incident electromagnetic field. Solve the rate equations.

2.5 An electromagnetic field of radiation frequency $6 \times 10^3\ GHz$ is incident on nearly resonant two level atoms under thermal equilibrium at a temperature T, such that $k_B T = 200\ cm^{-1}$. Find the relative population distribution in the two levels (assume Boltzmann distribution). Solve the rate equations under RWA at resonance and show that the population oscillates at the Rabi frequency.

2.6 A two level atom interacts with a strong electromagnetic field. Consider an initial condition such that all particles are in the upper state b. Solve the rate equations to obtain the transition probability. (Hint: Assume a trial solution $C_b(t) = A\, e^{ift}$).

2.7 A strong electromagnetic field interacts with a two-level atom such that the radiation frequency is resonant with the transition frequency of the atom. Find the time duration of a pulse when the population of the two levels will be the same. What is the pulse duration if the Rabi frequency is 1 MHz?

2.8 In problem 2.7, if the radiation is switched off after the pulse duration, when the two levels have equal population and the probability of spontaneous emission is negligible, such that over a time when the radiation pulse remains switched off and there is no spontaneous emission, what will be the result at the end of a pulse of same duration that is applied again?

2.9 A two level atom interacts with a strong electromagnetic field with the radiation frequency at resonance with the atomic transition frequency. The Rabi solutions are $f_1 = \Omega/2$ and $f_2 = -\Omega/2$. Show that the C – coefficients (Eq. 2.5.9 and 2.5.10) can be written as

$$C_a(t) = i \sin(\Omega t/2)$$

$$C_b(t) = \cos \Omega t/2$$

From equation for the total wave function (Eq.2.2.3), show that there are four oscillation frequencies at $\omega_a + \Omega/2$, $\omega_a - \Omega/2$, $\omega_b + \Omega/2$ and $\omega_b - \Omega/2$ (Fig. 2.8). With the help of energy level diagram show that each energy level is split into two levels. (This splitting is called Rabi or AC Stark splitting at strong field. The transition frequency at ω is split into three transitions at ω, $\omega \pm \Omega$.)

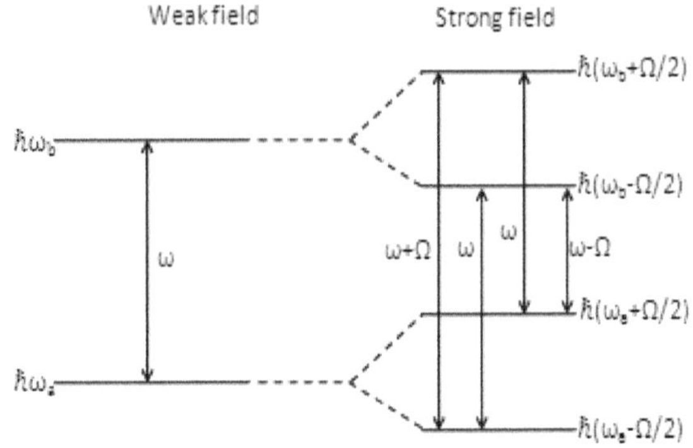

Fig. 2.8 Energy level splitting in presence of a strong field.

2.10 By adding phenomenological decay constants to the rate equations to the C-coefficients for interaction of a two-level atom with a strong electromagnetic field we get

$$\dot{C}_a(t) = i\frac{\Omega}{2}e^{-i(\omega-\nu)t}C_b(t) - \frac{1}{2}\gamma_a C_a,$$

$$\dot{C}_b(t) = i\frac{\Omega}{2}e^{i(\omega-\nu)t}C_b(t) - \frac{1}{2}\gamma_b C_b.$$

Use Rabi method of exact solution to obtain the eigenfrequencies.

$$f_{1,2} = -\frac{1}{2}[\omega - \nu - \frac{i}{2}(\gamma_a + \gamma_b)] \pm \frac{1}{2}[\{\omega - \nu - \frac{i}{2}(\gamma_b - \gamma_a)\}^2 + \Omega^2]^{1/2} .$$

Show that the complex Rabi frequency is

$$\mathrm{F} = [\{\omega - \nu - \frac{i}{2}(\gamma_b - \gamma_a)\}^2 + \Omega^2]^{1/2}$$

Calculate the oscillation time period in the resonant case.

2.11 If the average thermal velocity is 800 m/sec and the mean free path is 6 micron, calculate the time between collisions and the duration of the collision.

2.12 For sodium D_1 line at 589.0 nm, show that at T=500K the Doppler width is nearly 0.02Å.

Further reading

1. E. Merzbacher, Quantum Mechanics, John Wiley, New York, 1970.
2. S. Gasiorowicz, Quantum Physics, John Wiley, New York, 1974.
3. M. Sargent III, M.O. Scully and W.E Lamb Jr., Laser Physics, Addison-Wesley, London, 1974.
4. S. C. Rand, Non-linear and Quantum Optics, Oxford, New York, 2010.
5. S. Stenholm, Foundations of Laser Spectroscopy, Wiley-Interscience, New York, 1983.

Chapter 3
Density Matrix Equations

3.1 The Density matrix

The quantum mechanical treatment of atom describes a single atom. Hence, the atom field interaction discussed in the last chapter describes the interaction of a single atom with a classical electromagnetic field. In any experiment, under usual circumstances, we have a large number of atoms interacting with the field. The observed result is an average over the effect of interaction of the electromagnetic field with all these atoms. The atoms may not all be in the same quantum mechanical state. For the atom in one quantum mechanical state, the observed value is an expectation value which is an average over a large number of measurements of the same physical variable in the same state. The atom is described by a set of eigenstates φ_n. At any instant the general wave function of an atom is

$$|\psi\,(t)\rangle = \textstyle\sum_n c_n(t)\,|\phi_n\rangle. \tag{3.1.1}$$

where $\quad c_n(t) = C_n(t)e^{-i\omega_n t}$

is the Schrödinger picture representation describing the complete time dependence and $C_n(\mathrm{t})$ is the interaction picture time dependence.

The expectation value of a linear operator P can be written as

$$\langle P\rangle = \textstyle\sum_n\sum_m c_n^*\, c_m \int \varphi_n^*\, P\varphi_m d\tau$$

$$= \textstyle\sum_n\sum_m c_n^*\, c_m\,\langle\varphi_n|P|\varphi_m\,\rangle$$

$$= \textstyle\sum_n\sum_m c_n^*\, c_m\, P_{nm}, \tag{3.1.2}$$

where, $$P_{nm} = \langle\varphi_n|P|\varphi_m\rangle. \tag{3.1.3}$$

Hence, in a mixed state or in a linear combination of eigenstates, the observable quantity depends on the *bilinear combination* of the states of the atom. The bilinear combination $c_n^* c_m$ may be written in the form of a matrix ρ_{nm}. In the case of a two-level atom described in the previous chapter we can write

$$\rho = \begin{pmatrix} c_a c_a^* & c_a c_b^* \\ c_b c_a^* & c_b c_b^* \end{pmatrix}, \tag{3.1.4}$$

where, $\rho_{aa} = c_a c_a^*$ and $\rho_{bb} = c_b c_b^*$ are the probability of occupation of the lower and upper states respectively. $\rho_{ab} = c_a c_b^*$ and $\rho_{ba} = c_b c_a^*$ are proportional to the transition dipole moment, if the transition is allowed. The matrix ρ defined in Eq. (3.1.4) is called the **Density Matrix**.

Hence the expectation value of an operator P is

$$\langle P\rangle = \textstyle\sum_{n,m} \rho_{mn} P_{nm}$$

$$= Tr\,(\rho P)\,. \tag{3.1.5}$$

For electric dipolar transition, the expectation value of the dipole moment is

$$\langle\mu\rangle = Tr\,(\rho\mu)$$

$$= (\rho_{ab}\mu_{ba} + \rho_{ba}\mu_{ab}). \qquad (3.1.6)$$

The diagonal matrix elements of μ are zero since the atomic eigenfuctions have definite parity and the dipole operator has odd parity. It may be noted that the non-diagonal matrix elements of density matrix play an important role in the transition.

Projection operator

An element of the density matrix may be written as

$$\rho_{mn} = c_m c_n^*$$

$$= \langle\varphi_m|\psi\rangle\langle\psi|\phi_n\rangle$$

$$= \langle\phi_m|\rho_{op}|\phi_n\rangle,$$

where $$\rho_{op} = |\psi\rangle\langle\psi| \qquad (3.1.7)$$

is the projection operator and density matrix is a representation of this *dyadic operator* in the space of atomic eigenfunctions in this case.

3.2 Density matrix of a statistical ensemble

As stated in the last section all measurements are usually made on a statistical ensemble of atomic systems. All atoms may not be in identical quantum mechanical states. An observation is an ensemble average over the states of all particles. In statistical mechanics, we have to take ensemble average, since we do not know the state of the system being investigated. The state of the system may be described by the density matrix for the ensemble average.

The quantum mechanical states are expressed as linear combinations with c coefficients. But the set of c coefficients may differ for different atoms. Hence, we have to average over all such combinations of states. This i-th combination of eigenstates for one atom may be written as

$$|\psi_i\,(t)\rangle = \sum_n c_n^i(t)\,|\phi_n\rangle. \qquad (3.2.1)$$

We consider an N-particle system. For an atom in the state $|\psi_i\rangle$ the quantum mechanical expectation value is $\langle\psi_i|P|\psi_i\rangle$, but the statistical average over the ensemble is

$$\overline{\langle P\rangle} = \frac{1}{N}\sum_{i=1}^{N}\langle\psi_i|P|\psi_i\rangle. \qquad (3.2.2)$$

Since each state $|\psi_i\rangle$ can be expressed in the form of Eq. (3.2.1) in terms of the atomic eigenstates we can write

$$\overline{\langle P\rangle} = \frac{1}{N}\sum_{i=1}^{N}\sum_{n,m} c_n^{i*}\,c_m^i\langle\phi_n|P|\phi_m\rangle \quad = \sum_{n.m}\rho_{mn}\,P_{nm}$$

$$= \mathrm{Tr}\,(\rho P), \qquad (3.2.3)$$

where, $$\rho_{mn} = \frac{1}{N}\sum_{i=1}^{N} c_n^{i*} c_m^i = \overline{c_n^* c_m} \tag{3.2.4}$$

As shown in Eq. (3.1.7) the density projection operator for the statistical ensemble is

$$\rho_{op} = \overline{|\psi\rangle\langle\psi|} = \frac{1}{N}\sum_{i=1}^{N} |\psi_i\rangle\langle\psi_i|. \tag{3.2.5}$$

The equation of motion of the density operator is

$$\dot{\rho}_{op} = \frac{1}{N}\sum_{i=1}^{N}[|\dot{\psi}_i\rangle\langle\psi_i| + |\psi_i\rangle\langle\dot{\psi}_i|]$$

$$= -\frac{i}{\hbar}[H,\rho]. \tag{3.2.6}$$

Rate equations for the density matrix elements may be written as

$$\dot{\rho}_{jk} = -\frac{i}{\hbar}\langle j|(H\rho - \rho H)|k\rangle$$

$$= -\frac{i}{\hbar}\sum_l [\langle j|H|l\rangle\langle l|\rho|k\rangle - \langle j|\rho|l\rangle\langle l|H|k\rangle]$$

$$= -\frac{i}{\hbar}\sum_l (H_{jl}\rho_{lk} - \rho_{jl}H_{lk}). \tag{3.2.7}$$

This is the Heisenberg form of equation of motion. For treatment of interaction of atoms with fields these equations are useful.

Exercise 3.1 For a two-level atom in thermal equilibrium at temperature T, the density operator can be written as

$$\rho = \frac{1}{z}e^{-H/k_BT}$$

z is a normalization constant such that $\mathrm{Tr}(\rho) = 1$. Show that the diagonal element of ρ is exponentially decreasing with energy and the coherence between the levels is zero.

Solution

For a two level atom, the Hamiltonian is diagonal, hence we get for the diagonal elements of the density matrix

$$\rho_{ii} = \frac{1}{z}e^{-E_i/k_bT}$$

$$\rho_{ij} = 0. \quad \text{if } i \neq j.$$

3.3 Decay phenomena and Density matrix equations for two-level atoms

Using Eq. 2.1.8 and 2.1.9 for a two level atom (Fig. 3.1) and the definition of c coefficients we can write the rate equations for the c coefficients as

$$\dot{c}_a(t) = -i\omega_a c_a - \frac{i}{\hbar}V_{ba}c_b, \tag{3.3.1}$$

$$\dot{c}_b(t) = -i\omega_b c_b - \frac{i}{\hbar}V_{ab}c_a, \tag{3.3.2}$$

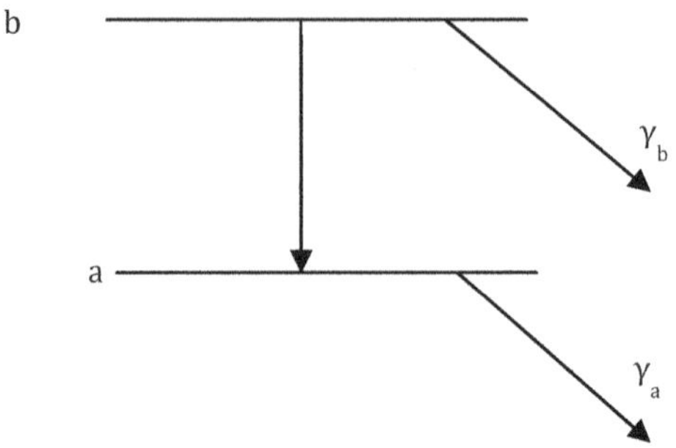

Fig. 3.1 Two level atom with population decay

Atoms in an excited level can decay in time, because of spontaneous emission. They can also decay from all the levels by other effects, such as collisions and other phenomena in the atomic media. The decay due to spontaneous emission will be discussed later by quantum treatment of radiation (Chapter 7). In general, we can include all the decay effects by introducing *phenomenological decay constants.*

$$\dot{c}_a(t) = -(i\omega_a + \tfrac{1}{2}\gamma_a)\, c_a - \tfrac{i}{\hbar}\, V_{ab} c_b, \tag{3.3.3}$$

$$\dot{c}_b(t) = -\left(i\omega_b + \tfrac{1}{2}\gamma_b\right) c_b - \tfrac{i}{\hbar}\, V_{ba} c_a. \tag{3.3.4}$$

These equations show that the probability of occupation of a level a in absence of the applied field is given by

$$|c_a|^2 = e^{-\gamma_a t}. \tag{3.3.5}$$

The rate equations for the c-coefficients lead to the equations for the density matrix components.

$$\dot{\rho}_{ab} = \tfrac{1}{N}\textstyle\sum_i [\, \dot{c}_a^i\, c_b^{i*} + c_a^i \dot{c}_b^{i*}] = \tfrac{1}{N}\sum_i [\, -i(\omega_a - \omega_b) c_a^i\, c_b^{i*} - \tfrac{i}{\hbar} V_{ab}(c_b^i c_b^{i*} - c_a^i c_a^{i*}) - \tfrac{\gamma_a}{2} c_a^i c_b^{i*} - \tfrac{\gamma_b}{2} c_a^i c_b^{i*}\,]$$

$$= i\omega\rho_{ab} - \tfrac{i}{\hbar} V_{ab}(\rho_{bb} - \rho_{aa}) - \gamma_{ab}\rho_{ab}, \tag{3.3.6}$$

where, $\gamma_{ab} = (\gamma_a + \gamma_b)/2.$

Similarly, the equations for the other density matrix components are

$$\dot{\rho}_{aa} = \left(-\tfrac{i}{\hbar} V_{ab}\rho_{ba} + C.C.\right) - \Upsilon_a \rho_{aa}, \tag{3.3.7}$$

$$\dot{\rho}_{bb} = \left(-\tfrac{i}{\hbar} V_{ba}\rho_{ab} + C.C.\right) - \Upsilon_b \rho_{bb}, \tag{3.3.8}$$

and

$$\dot{\rho}_{ba} = -i\omega\rho_{ba} + \tfrac{i}{\hbar} V_{ba}(\rho_{bb} - \rho_{aa}) - \gamma_{ab}\rho_{ba}. \tag{3.3.9}$$

The first term in equations for the non-diagonal elements arises from the definition of c coefficients and will not appear in the interaction picture. This term leads to rapid variation at frequency ω.

The decay terms introduced *phenomenologically* in this section play an important role in the operation of laser. At microwave frequency the lifetime in the states is long compared to the Rabi oscillation period and the transit time. At optical frequencies, the spontaneous emission from upper states is important and the lifetime is small. In such cases, the lifetime for emission from the lower level is important, so that atoms do not start accumulating in this state. The decay of the coherence term arises from elastic collisions in a gas. This term is important in the laser operation, as well in the case of spectroscopy. The decay in of the non-diagonal terms of the density matrix may vary independent of the level decay rates. The rate equations derived from the equations for the c-coefficients, of course, depend on the level decay rates.

3.4 Vector model solution of density matrix equations

The set of three density matrix equations can be solved simultaneously by direct integration method and by using the initial conditions. Eq. (3.3.9) is complex conjugate of Eq. (3.3.6). For spectral dynamics, it is important to know the time behaviour of the off-diagonal density matrix elements. We can show that a redefinition of the three elements can lead to a vector representation and the equations can be cast into a vector equation that can be compared with a precessing magnetic dipole. A two-level atom is similar to a spin-1/2 magnetic dipole. The equations are comparable to the **Bloch equations** used in Nuclear Magnetic Resonance. This is only an analogy. There are differences in the decay constants as we have three constants in this case. This situation exists only in some laser media. The equations (3.3.7-9) are referred as **Optical Bloch Equations** when the interacting field is an electromagnetic radiation field.

Under Rotating Wave Approximation (RWA), the atom field interaction energy for the transition dipole interaction case can be written as

$$V_{ba} = -\frac{1}{2}\mu\mathcal{E}_0 e^{-ivt}. \qquad (3.4.1)$$

Multiplying Eq. (3.3.9) by e^{ivt} we get

$$\dot{\rho}_{ba}e^{ivt} = -(i\omega + \gamma)\rho_{ba}e^{ivt} - \frac{i}{2\hbar}\mu\mathcal{E}_0(\rho_{bb} - \rho_{aa}).$$

We have assumed $\gamma_{ab} = \gamma$

$$\frac{d}{dt}\left(\rho_{ba}e^{ivt}\right) = -[i(\omega - v) + \gamma]\rho_{ba}e^{ivt} -$$

$$\frac{i}{2\hbar}\mu\mathcal{E}_0(\rho_{bb} - \rho_{aa}). \qquad (3.4.2)$$

Introducing the variables

$$R_1 = \rho_{ba}e^{ivt} + C.C. \qquad (3.4.3)$$

$$R_2 = i\rho_{ba}e^{ivt} + C.C, \qquad (3.4.4)$$

and

$$R_3 = \rho_{bb} - \rho_{aa} \qquad (3.4.5)$$

we get

$$R_1 - iR_2 = 2\rho_{ba}e^{ivt}. \qquad (3.4.6)$$

From Eq. (3.4.2) we get

$\frac{d}{dt}(R_1 - iR_2) = = -[i(\omega - \nu) + \gamma](R_1 - iR_2) - \frac{i}{\hbar}\mu\mathcal{E}_0 R_3.$

By comparing the coefficients of the real and imaginary parts we get, after defining $\gamma = \frac{1}{T_2}$

$$\dot{R}_1 = -\frac{1}{T_2}R_1 - (\omega - \nu)R_2 , \tag{3.4.7}$$

$$\dot{R}_2 = -\frac{1}{T_2}R_2 + (\omega - \nu)R_1 + \frac{1}{\hbar}\mu\mathcal{E}_0 R_3. \tag{3.4.8}$$

Using $\gamma_a = \gamma_b = \frac{1}{T_1}$, we can write

$$\dot{R}_3 = \dot{\rho}_{bb} - \dot{\rho}_{aa}$$

$$= -\frac{1}{T_1}(\rho_{bb} - \rho_{aa}) - (\frac{2i}{\hbar}V_{ba}\rho_{ab} + C.C.)$$

$$= -\frac{1}{T_1}(\rho_{bb} - \rho_{aa}) + (\frac{i}{\hbar}\mu\mathcal{E}_0 e^{-i\nu t}\rho_{ab} + C.C.)$$

$$= -\frac{1}{T_1}R_3 - \frac{\mu\mathcal{E}_0}{\hbar}R_2. \tag{3.4.9}$$

It may be noticed that in this case, we have assumed all the decay constants to be equal to γ and they are same. This is a gross assumption and it may not be true in all cases. The set of equations are equivalent to the Bloch equations used in NMR, where the level decay constants are written as $1/T_1$ and the coherence decay constants are written as $1/T_2$, the relaxation times T_1 and T_2 are known as **Longitudinal and Transverse relaxation times**. The three variables may be considered as the components of a vector and we can then constitute a vector $\vec{R} = \vec{n}_1 R_1 + \vec{n}_2 R_2 + \vec{n}_3 R_3$ called the **Bloch vector**. The three equations 3.3.15-17 of the density matrix components may be written in a vector form if we define a vector

$$\vec{\mathcal{F}} = \frac{\mu\mathcal{E}_0}{\hbar}\vec{n}_1 - (\omega - \nu)\vec{n}_3 , \tag{3.4.10}$$

where $\Omega = \frac{\mu\mathcal{E}_0}{\hbar}$, defined by Eq. (2.2.5), is the Rabi frequency and $(\omega - \nu)$ is the detuning. If T_1 and T_2 are equal and we write

$$T_1 = T_2 = 1/\gamma.$$

The vector equation can be written as

$$\dot{\vec{R}} = -\gamma\vec{R} + \vec{R} \, X \, \vec{\mathcal{F}} . \tag{3.4.11}$$

The rate equation for $\vec{R}$ shows the precession of the vector $\vec{R}$ (Fig. 3.2) around the direction of $\vec{\mathcal{F}}$. The **precession frequency** has a magnitude $\mathcal{F}$ (Eq. 3.4.10), determined by the Rabi frequency and the detuning. The amplitude of the precessing vector diminishes at the rate γ.

$$\mathcal{F} = [(\omega - \nu)^2 + \Omega^2]^{1/2} . \tag{3.4.12}$$

At resonance, $(\omega - \nu) = 0$, hence the precession frequency is the Rabi frequency and the vector rotates around the direction $\vec{n}_1$.

If at $t = 0$, $R_1 = R_2 = 0$, $R_3 = 1$, we have a case of $\rho_{bb} = 1$ and $\rho_{aa} = 0$ and also $\rho_{ab} = \rho_{ba} = 0$. This is a situation with the atom at the upper level and the vector is pointing in the $+n_3$ direction. When the vector starts the precession, it will point in the downward or $-n_3$ direction after half period of oscillation i.e. at time $t_{1/2} = \frac{\pi}{\Omega}$. In this case, $R_3 = -1$ and $\rho_{bb} = 0$ and $\rho_{aa} = 1$. Hence, the atom is in the lower state. Thus, there is a transition from the upper level to the lower level. This can be shown pictorially (Fig. 3.2). *Thus, the system exhibits oscillation between the two levels with a frequency equal to the Rabi frequency.* A pulse of duration $\frac{\pi}{\Omega}$ is called **π – pulse**. Because of the decay terms, the rotating vector reduces in size and spirals down to the centre eventually. In the non-resonance case the vector will precess around a direction in the plane n_1, n_3 as shown on the right hand side of Fig. 3.2. In such cases, the vector will precess with frequency

$[(\omega - \nu)^2 + (\frac{\mu\mathcal{E}_0}{\hbar})^2]^{1/2}$ and the complete transition from upper to lower state (or inversion) never takes place.

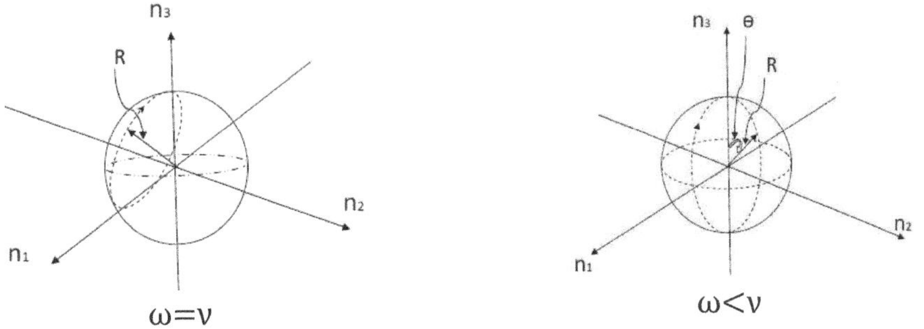

Fig. 3.2 Precession of the Bloch Vector. In the resonance case. the vector $\vec{R}$ precesses around the $\vec{n}_1$ direction with an angular frequency Ω. In the off-resonance case precession is around a direction in the $\vec{n}_1, \vec{n}_3$ plane with a frequency determined by the Rabi frequency and detuning

The vector model of the density matrix equations illustrates the atomic transitions in the two level atoms in the presence of a near resonant electromagnetic field in a very elegant manner. The description of atom field interaction by a vector model is only analogous to the case of spin-1/2 system in nuclear magnetic resonance. But there is a difference. In a spin system there is a rotation of the spin vector in real space around a direction fixed by the presence of a static magnetic field. But in the optical dipole case, there is no static field and there is no precession in the real space. In fact, there is no fixed direction associated with the population difference. But the system shows polarization. Hence the decay time associated with the decay of the transverse component of the vector is called the transverse decay time. The other decay time is called the longitudinal decay time.

3.5 Moving atoms in a progressive wave

We consider a two-level atom moving with a constant uniform velocity v in one direction z. In this case, the electric field of the electromagnetic field can be written as

$$\mathcal{E}(z,t) = \mathcal{E}_0 \cos(kz - \nu t)\,. \quad (3.5.1)$$

The density matrix equations for a stationary atom remain the same as Eq. (3.3.7–9), where the time derivative $\frac{d}{dt}$ is to be replaced by $\frac{\partial}{\partial t} + v\,\frac{\partial}{\partial z}$.

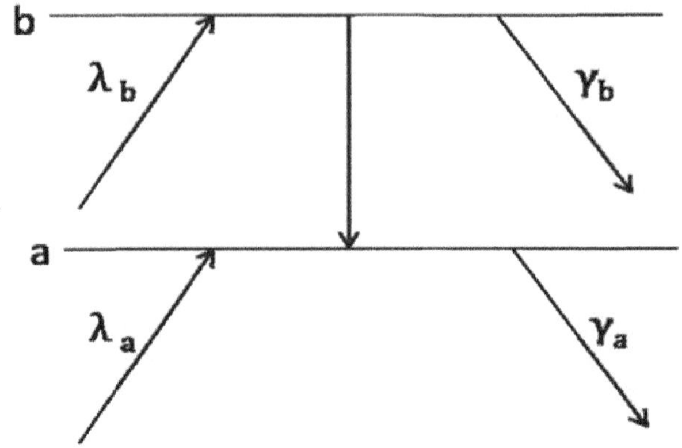

Fig. 3.3 Two-level atom showing pumping and population decay

Hence the equations are

$$(\frac{\partial}{\partial t}+v\ \frac{\partial}{\partial z})\ \rho_{aa}+\gamma_a\rho_{aa}=\ \lambda_a\ +i\frac{\mu\mathcal{E}_0}{\hbar}\cos(kz-\ v\ t)\ (\rho_{ba}-\rho_{ab}), \tag{3.5.2}$$

$$(\frac{\partial}{\partial t}+v\ \frac{\partial}{\partial z})\ \rho_{bb}+\gamma_b\rho_{bb}=\ \lambda_b-i\frac{\mu\mathcal{E}_0}{\hbar}\cos(kz-\ v\ t)\ (\rho_{ba}\ \text{-}\rho_{ab}) \tag{3.5.3}$$

and

$$(\frac{\partial}{\partial t}+v\ \frac{\partial}{\partial z}+i\omega)\rho_{ba}+\gamma_{ba}\rho_{ba}\ =\ -i\frac{\mu\mathcal{E}_0}{\hbar}\cos(kz-\ v\ t)\ (\rho_{bb}-\rho_{aa}). \tag{3.5.4}$$

We have assumed the pumping rates per unit time to the levels as λ_a and λ_b (Fig. 3.3). By assuming the transformation

$$\rho_{ba}=\ \tilde{\rho}_{ba}e^{i\,(\,kz-\nu t\,)}, \tag{3.5.5}$$

the derivative term for ρ_{ba} can be written as

$$(\frac{\partial}{\partial t}+v\frac{\partial}{\partial z})\ \rho_{ba}=-i(\nu-kv)\rho_{ba}. \tag{3.5.6}$$

From RWA, we can show

$$e^{\pm i\,(\,kz-\nu t\,)}\cos(kz-\ \nu\ t)=1/2\,.$$

From Eq. (3.5.4), we can write

$$[\gamma_{ba}+\ i\omega\ \ -i(\nu-kv)]\ \rho_{ba}=\ -i\frac{\mu\mathcal{E}_0}{\hbar}\cos(kz-\ \nu\ t).\ \ (\rho_{bb}-\ \rho_{aa}). \tag{3.5.7}$$

Multiplying both sides by $e^{-\,i\,(\,kz-\nu t\,)}$, we get

$$\tilde{\rho}_{ba}\ =\ -i\frac{\mu\mathcal{E}_0}{2\hbar}\ \frac{(\rho_{bb}-\rho_{aa})}{[i(\,\Delta+kv)+\gamma_{ba}\,]}\ . \tag{3.5.8}$$

For obtaining the steady state population difference, we can put the time derivatives of the population matrix elements equal to zero. Hence,

$$\gamma_a \rho_{aa} = \lambda_a + i\frac{\mu\mathcal{E}_0}{\hbar}\cos(kz - \nu t)(\rho_{ba} - \rho_{ab}), \tag{3.5.9}$$

$$\gamma_b \rho_{bb} = \lambda_b - i\frac{\mu\mathcal{E}_0}{\hbar}\cos(kz - \nu t)(\rho_{ba} - \rho_{ab}). \tag{3.5.10}$$

We can show that

$$\cos(kz - \nu t)(\rho_{ba} - \rho_{ab}) = \frac{1}{2}[e^{i(kz-vt)} + e^{-i(kz-vt)}][\tilde{\rho}_{ba}e^{i(kz-vt)} - \tilde{\rho}_{ab}e^{-i(kz-vt)}]$$

$$= \frac{1}{2}(\tilde{\rho}_{ba} - \tilde{\rho}_{ab})$$

$$= -\frac{1}{2}\frac{\mu\mathcal{E}_0}{2\hbar}(\rho_{bb} - \rho_{aa})[\frac{i}{i(\Delta+kv)+\gamma_{ba}} - \frac{-i}{-i(\Delta+kv)+\gamma_{ba}}]$$

$$= -i\frac{\mu\mathcal{E}_0}{4\hbar}(\rho_{bb} - \rho_{aa})\frac{2\gamma_{ba}}{\gamma_{ba}^2 + (\Delta+kv)^2}$$

$$= -i\frac{\mu\mathcal{E}_0}{2\hbar}(\rho_{bb} - \rho_{aa})\frac{1}{\gamma_{ba}}L(\Delta + kv), \tag{3.5.11}$$

where

$$L(\Delta + kv) = \frac{\gamma_{ba}^2}{\gamma_{ba}^2 + (\Delta+kv)^2},$$

$L(\Delta)$ is a dimensionless Lorentzian.

After introducing Eq. (3.5.11) into Eq. (3,5.9-10) we have

$$\gamma_a \rho_{aa} = \lambda_a + i\frac{\mu\mathcal{E}_0}{\hbar}(-i\frac{\mu\mathcal{E}_0}{2\hbar})(\rho_{bb} - \rho_{aa})\frac{1}{\gamma_{ba}}L(\Delta + kv), \tag{3.5.12}$$

$$\gamma_b \rho_{bb} = \lambda_b - i\frac{\mu\mathcal{E}_0}{\hbar}(-i\frac{\mu\mathcal{E}_0}{2\hbar})(\rho_{bb} - \rho_{aa})\frac{1}{\gamma_{ba}}L(\Delta + kv), \tag{3.5.13}$$

$$(\rho_{bb} - \rho_{aa}) = \frac{\lambda_b}{\gamma_b} - \frac{\lambda_a}{\gamma_a} - (\frac{\mu\mathcal{E}_0}{\hbar})^2\frac{1}{2\gamma_{ba}}L(\Delta + kv)(\frac{1}{\gamma_b} + \frac{1}{\gamma_a})(\rho_{bb} - \rho_{aa}), \tag{3.5.14}$$

or

$$(\rho_{bb} - \rho_{aa}) = \frac{\frac{\lambda_b}{\gamma_b} - \frac{\lambda_a}{\gamma_a}}{1 + I\frac{\gamma_b + \gamma_a}{2\gamma_{ba}}L(\Delta+kv)} = \frac{\bar{N}}{1+2I\eta L(\Delta+kv)}, \tag{3.5.15}$$

$$\bar{N} = \frac{\lambda_b}{\gamma_b} - \frac{\lambda_a}{\gamma_a}, \tag{3.5.16}$$

$$I = (\frac{\mu\mathcal{E}_0}{\hbar})^2\frac{1}{2\gamma_a\gamma_b}, \tag{3.5.17}$$

$$\eta = \frac{\gamma_b + \gamma_a}{2\gamma_{ba}}. \tag{3.5.18}$$

$\bar{N}$ is the difference between probabilities of occupation between the two levels when there is no field, or it may be called the **unperturbed population difference,** I is the **dimensionless field intensity** and η is the **saturation decay parameter**. The Lorentzian function leads to a resonance condition, when $\Delta = -kv$ or $\nu = \omega + kv$. This arises from Doppler shift. Since the *unsaturated population difference is a function of velocity* we can write

$$(\rho_{bb} - \rho_{aa}) = \bar{N}(v)/[1 + \frac{2I\eta\gamma_{ba}^2}{\gamma_{ba}^2 + (\Delta+kv)^2}]$$

$$= \bar{N}(v)(\frac{\gamma_{ba}^2 + (\Delta+kv)^2 + 2I\eta\gamma_{ba}^2}{\gamma_{ba}^2 + (\Delta+kv)^2})^{-1}$$

$$= \bar{N}(v)[1 - \frac{2I\eta\gamma_{ba}^2}{\gamma_{ba}^2 + (\Delta+kv)^2 + 2I\eta\gamma_{ba}^2}]. \quad (3.5.19)$$

Hence the population difference is a deviation from the unperturbed population difference measured by a Lorentzian centred at $\nu = \omega + kv$ and having a power broadened width

$$\gamma_p = \gamma_{ba}(1 + 2I\eta)^{1/2}. \quad (3.5.20)$$

The additional intensity dependent term is responsible for power or **saturation broadening**.

Hence, the population difference curve plotted against velocity will show a hole as shown in Fig. 3.4. This hole is called the **Bennett hole**. The physical explanation of the appearance of such a hole is the lower population difference caused by the transition to the upper state of a particular velocity group of atoms that resonate with the radiation frequency. The noticeably sharp feature is explained by the lower value of γ_p. It is possible because of the high monochromaticity of the laser radiation. The hole shows a Doppler shift and the Lorentzian hole on the Doppler background is power broadened. From Eq. (3.5.12) and (3.5.13) we get

$$(\rho_{bb} + \rho_{aa}) = \frac{\lambda_b}{\gamma_b} + \frac{\lambda_a}{\gamma_a},$$

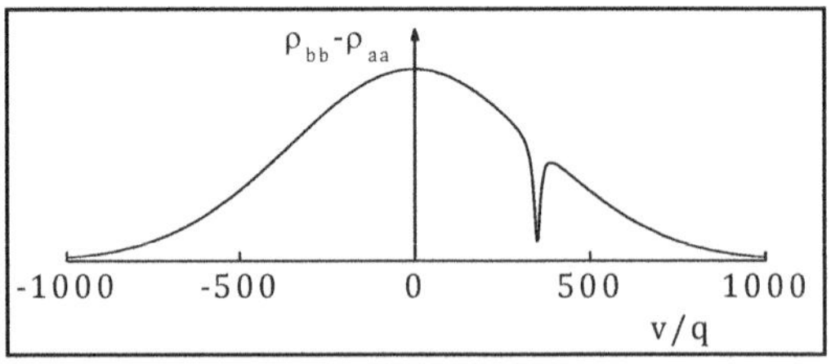

Fig. 3.4 A hole burnt on the population difference curve.

and it can be shown that

$$\rho_{bb} = \frac{\lambda_b}{\gamma_b} - \bar{N}(v)\frac{2I\eta\gamma_{ba}^2}{\gamma_{ba}^2 + (\Delta+kv)^2 + 2I\eta\gamma_{ba}^2}, \quad (3.5.21)$$

$$\rho_{aa} = \frac{\lambda_a}{\gamma_a} + \bar{N}(v)\frac{2I\eta\gamma_{ba}^2}{\gamma_{ba}^2 + (\Delta+kv)^2 + 2I\eta\gamma_{ba}^2}. \quad (3.5.22)$$

Thus the lower state shows a Lorentzian hole, while the upper state exhibits a Lorentzian peak (Fig. 3.5)

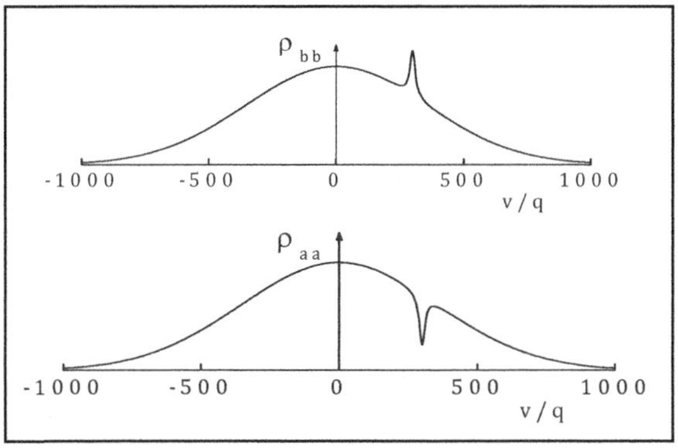

Fig. 3.5 The hole and the peak in the population of the lower and upper levels.

The imaginary part of $\tilde{\rho}_{ba}$ from Eq. (3.5.8) and (3.5.15) is

$$Im\,\tilde{\rho}_{ba} = \frac{1}{2i}\left(\tilde{\rho}_{ba} - \tilde{\rho}_{ab}\right)$$

$$= \frac{i\mu\mathcal{E}_0}{4\hbar}\left(\rho_{bb} - \rho_{aa}\right)\left[\frac{i}{i(\Delta+kv)+\gamma_{ba}} - \frac{-i}{-i(\Delta+kv)+\gamma_{ba}}\right]$$

$$= -\frac{\mu\mathcal{E}_0}{2\hbar}\left(\rho_{bb} - \rho_{aa}\right)\frac{\gamma_{ba}}{\gamma_{ba}{}^2+(\Delta+kv)^2}$$

$$= -\frac{\mu\mathcal{E}_0}{2\hbar}\frac{\bar{N}}{1+2I\eta L\,(\Delta+kv)}\,\frac{\gamma_{ba}}{\gamma_{ba}{}^2+(\Delta+kv)^2}$$

$$= -\frac{\mu\mathcal{E}_0}{2\hbar}\frac{\bar{N}\gamma_{ba}}{[\gamma_{ba}{}^2+(\Delta+kv)^2][\,1+2I\eta\frac{\gamma_{ba}^2}{\gamma_{ba}^2+(\Delta+kv)^2}]}$$

$$= -\frac{\mu\mathcal{E}_0}{2\hbar}\frac{\bar{N}\gamma_{ba}}{(\Delta+kv)^2+\gamma_{ba}^2(1+2I\eta)}. \qquad (3.5.23)$$

This is the power broadened response from all atoms with a particular velocity v. The power broadened width is the same as in Eq.(3.5.20).

In order to calculate the total polarization we have to integrate $\tilde{\rho}_{ba}$ given in equation (3.5.8) over all velocities, as the density matrix elements are functions of velocity.

Total polarization can be written as

$$P(z,t) = N_0\mu \int [\,\rho_{ba}(v,z,t) + \rho_{ab}(v,z,t)]\,dv. \qquad (3.5.24)$$

Note that ρ_{ba} and ρ_{ab} are functions of z, t defined by the transformation of Eq. (3.5.5). N_0 is the number of atoms per unit volume. As shown above they are also functions of velocity for the moving atoms. In terms of $\tilde{\rho}_{ba}$ we can write

$$P(z,t) = N_0\mu \int [\,\tilde{\rho}_{ba}(v)e^{i(kz-\nu t)} + \tilde{\rho}_{ab}(v)\,e^{-i(kz-\nu t)}]\,dv. \qquad (3.5.25)$$

In terms of the complex susceptibility we can also write polarization as

$$P(z,t) = \frac{1}{2}\epsilon_0\mathcal{E}_0\left[\chi e^{i(kz-\nu t)} + \chi^* e^{-i(kz-\nu t)}\right]. \tag{3.5.26}$$

From these two equations, we get

$$\chi = \frac{2N_0\mu}{\epsilon_0\mathcal{E}_0}\int \tilde{\rho}_{ba}(v)\,dv$$

$$= \frac{2N_0\mu}{\epsilon_0\mathcal{E}_0}\int\left(-i\frac{\mu\mathcal{E}_0}{2\hbar}\right)\frac{(\rho_{bb}-\rho_{aa})}{[i(\Delta+kv)+\gamma_{ba}]}\,dv$$

$$= -i\frac{N_0\mu^2}{\epsilon_0\hbar}\int\frac{\bar{N}(v)}{1+2I\eta L(\Delta+kv)}\frac{\gamma_{ba}-i(\Delta+kv)}{\gamma_{ba}^2+(\Delta+kv)^2}\,dv,$$

where we have used Eq.(3.5.8) and Eq. (3.5.15). Substituting the expression for the dimensionless Lorentzian $L(\Delta+kv)$, we have

$$\chi = -i\frac{N_0\mu^2}{\epsilon_0\hbar}\int\frac{\gamma_{ba}-i(\Delta+kv)}{(\Delta+kv)^2+\gamma_{ba}^2(1+2I\eta)}\bar{N}(v)dv. \tag{3.5.27}$$

The unsaturated population difference $\bar{N}(v)$ obeys the Maxwell-Boltzmann velocity distribution law.

$$\bar{N}(v) = \frac{1}{\sqrt{\pi}q}\bar{N}_0 e^{-\frac{v^2}{q^2}}, \tag{3.5.28}$$

where

$$q = \sqrt{\frac{2k_BT}{M}}, \tag{3.5.29}$$

q is 1/e of the width the velocity distribution, k_B is Boltzmann constant, T is temperature and M is the atomic mass.

Taking the imaginary part of susceptibility we get

$$Im(\chi) = \frac{N_0\mu^2}{\epsilon_0\hbar\sqrt{\pi}q}\int\frac{\gamma_{ba}}{(\Delta+kv)^2+\gamma_{ba}^2(1+2I\eta)}\bar{N}_0\, e^{-\frac{v^2}{q^2}}dv. \tag{3.5.30}$$

Hence the loss or gain in the medium is given by a convolution of a Lorentzian and a Gaussian function, called a **Voigt profile**. The integration will lead to the same spectral profile as in Fig. 3.4.

The Gaussian line width is the **Inhomogeneous Line width**, it varies from atom to atom as the velocities differ, and line width of the Lorentzian is called the **Homogeneous Line width** that applies to all atoms in the system. In the usual cases at normal or higher temperature, the *inhomogeneous width kq far exceeds the homogeneous width* γ_{ba}. The plasma dispersion function is given by Eq. (3.5.8). From Eq. (3.5.30) we can find the **absorption** part

$$Im(\chi) = \frac{N_0\mu^2}{\epsilon_0\hbar\sqrt{\pi}q}\frac{\bar{N}_0 e^{-\frac{\Delta^2}{k^2q^2}}}{(1+2I\eta)^{1/2}}\int\frac{\gamma_p}{(\Delta+kv)^2+\gamma_p^2}\,dv$$

$$= \frac{\sqrt{\pi}N_0\bar{N}_0\mu^2 e^{-\frac{\Delta^2}{k^2q^2}}}{\epsilon_0\hbar kq(1+2I\eta)^{1/2}}. \tag{3.5.31}$$

This is the Doppler limit and permits asymptotic evaluation. In a similar way, we can evaluate the **dispersive** part from the real part of the susceptibility as

$$Re\,(\chi\,) = -\frac{N_0\bar{N}_0\mu^2}{\epsilon_0\hbar\sqrt{\pi}q}\int\frac{\Delta+kv}{(\,\Delta+kv)^2+\gamma_p^2}e^{-\frac{v^2}{q^2}}dv. \quad (3.5.32)$$

The integral in the form $\int\frac{x}{x^2+\gamma_p^2}dv$ cannot be evaluated in a convergent form as easily as in the earlier case.

From the above we can conclude about the intensity dependence of susceptibility.

When $I \rightarrow 0$, both the real and imaginary parts are constant. So the polarization is linear in the electric field intensity $\mathcal{E}$, or $P = \chi(0)\,\mathcal{E}$. This phenomenon is observed in conventional **linear spectroscopy**.

When intensity $I \gg 1$, the imaginary part of $\chi \propto I^{-1/2}$ or $\chi \propto 1/\,\mathcal{E}$. The real part remains constant. Thus the power broadening affects only the absorption (or gain) part of the interaction. This leads to **nonlinear terms**. In general the intensity dependence leads to a power series in I or in $\mathcal{E}$.

Problems

3.1 The wave function of a two-level atom is given by

$$|\psi(t)\rangle = \sum_{n=a,b} C_n(t)e^{-i\omega_n t}\,|\phi_n\rangle$$

Find the expectation value of an operator P in the state $|\psi(t)\rangle$ in terms of the C-coefficients. Write down the expectation value of the dipole moment operator.

3.2 (a)Write down the density matrix of an unperturbed two-level atom where the occupation probability of the ground state is 0.75.

(b) Show that the non-diagonal matrix element of the projection operator $|\psi\rangle\langle\psi|$ is zero, if $|\psi\rangle$ is an eigenstate, but it is non-zero, if $|\psi\rangle = c_a|a\rangle + c_b|b\rangle$, a superposition of two states.

3.3 Show that for a pure state

$$\rho^\dagger = \rho,$$

$$\mathrm{Tr}\,(\rho) = 1,$$

$$\rho^2 = \rho,$$

$$\mathrm{Tr}\,(\rho^2) = 1.$$

Are the last two equations also true for the mixed state? Justify.

3.4 Use the generalized form of density matrix equation for the components of a multilevel atom

$$\dot{\rho}_{jk} = -\frac{i}{\hbar}\sum_l (H_{jl}\rho_{lk} - \rho_{jl}H_{lk})$$

to obtain the following equations for the components of the density matrix of a two-level atom interacting with an electromagnetic field.

$$\dot{\rho}_{aa} = \left(\frac{i}{2\hbar}\mu\mathcal{E}_0 e^{ivt}\rho_{ba} + C.C.\right)$$

$$\dot{\rho}_{bb} = \left(\frac{i}{2\hbar}\mu\mathcal{E}_0 e^{-ivt}\rho_{ab} + C.C.\right)$$

$$\dot{\rho}_{ba} = i\omega\rho_{ba} - \frac{i}{2\hbar}\mu\mathcal{E}_0 e^{-ivt}(\rho_{bb} - \rho_{aa})$$

$$\dot{\rho}_{ab} = -i\omega\rho_{ab} + \frac{i}{2\hbar}\mu\mathcal{E}_0 e^{ivt}(\rho_{bb} - \rho_{aa})$$

Compare them with the equations (3.3.6-9). Add phenomenological decay constants.

3.5 In the resonance case of interaction of a two level atom with a radiation field, show that the population difference or the component 3 of the Bloch vector shows sinusoidal oscillation, if the decay constants are zero. In the case of non-zero decay and zero detuning, it will exhibit damped oscillation. Write down the equations of motion in the two cases and give physical explanation.

3.6 Obtain the rate equations for the components of the Bloch vector without using RWA, i.e.

$$V_{ba} = -\frac{1}{2}\mu\mathcal{E}_0\left(e^{-i\nu t} + C.C.\right).$$

3.7 Consider the free evolution of a two-level system interacting with a radiation field detuned from the resonance frequency. If the field is switched off i.e. $\mathcal{E}_0 = 0$, show that the time evolution of the vector components R_1 and R_2 can be represented by a rotation matrix that shows sinusoidal oscillation with exponential decay

$$\begin{pmatrix} R_1(t) \\ R_2(t) \end{pmatrix} = e^{-t/T_2}\begin{pmatrix} \cos\Delta t & -\sin\Delta t \\ \sin\Delta t & \cos\Delta t \end{pmatrix}\begin{pmatrix} R_1(0) \\ R_2(0) \end{pmatrix}$$

and the population difference decays exponentially

$$R_3(t) = R_3(0)\, e^{-t/T_1}.$$

Discuss physical interpretation in terms of free induction decay.

3.8 A two level atom is moving with a velocity v and interacts with an electromagnetic field that is detuned from the resonance frequency by Δ. Find the condition when the ratio of the unsaturated population difference $(\rho_{bb} - \rho_{aa})$ and the saturated or unperturbed population difference $\overline{N}(v)$ is ½ for the resonant case $\Delta = -kv$.

3.9 Expand the imaginary part of susceptibility in a power series of Rabi frequency. Obtain the relation between the Rabi frequency and decay constants when the series converges. Show that the first non-linear term in susceptibility is third order in electric field amplitude.

3.10 Consider a three level atom a, b, c (Fig. 3.6) with energy levels $E_c > E_b > E_a$.

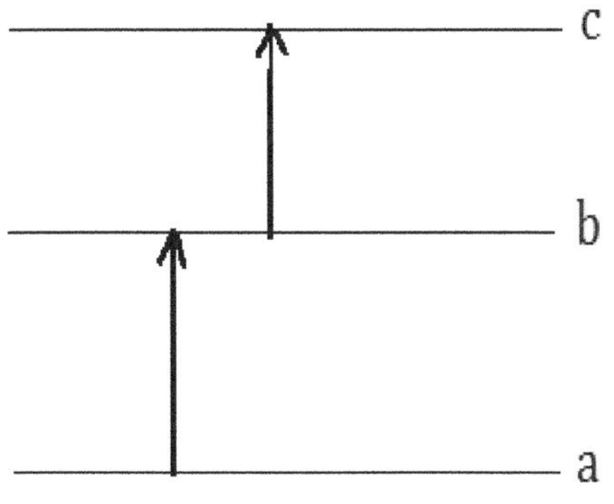

Fig. 3.6 A three level atom energy diagram

The atom interacts with two electromagnetic radiation fields resonant respectively with energy level differences $E_b - E_a$ and $E_c - E_b$.

Set up the 3x3 density matrix for the interacting system. Assume the transitions between the levels to be allowed.

3.11 In the problem of 3.10 consider the radiation field to be resonant with the energy level differences (i) $E_b - E_a$ and $E_c - E_a$, and (ii) $E_c - E_a$ and $E_c - E_b$. Set up the density matrix elements and draw the energy level diagrams.

Further reading

1. M. Sargent III, M.O. Scully and W.E Lamb Jr., Laser Physics, Addison-Wesley, London, 1974.
2. S. C. Rand, Non-linear and Quantum Optics, Oxford, New York, 2010.
3. Stenholm, Foundations of Laser Spectroscopy, Wiley-Interscience, New York, 1983.
4. P. Meystre and M. Sargent III, Elements of Quantum Optics, Springer, Berlin, 2007.
5. L. Allen and J. H. Eberly, Optical Resonance and Two-level Atoms, Dover, New York, 1987.

Chapter 4
Saturation Absorption Spectroscopy

4.1 Moving atoms in a standing wave

In the preceding chapter, we discussed the interaction of a moving atom with a progressive electromagnetic wave. It was shown that the velocity distribution leads to a Gaussian profile. But a relatively strong field may also cause Bennett Hole in the population distribution curve and that has a Lorentzian line shape. The position of the hole is shifted from the centre of the Gaussian curve by the amount of off-resonance from the laser frequency. In this chapter, we consider the radiation confined in a one dimensional cavity with two reflecting mirrors at the two ends of the cell of length L. The moving atoms will interact with the two oppositely moving waves and the interaction is expected to lead to a hole by each wave depending on the detuning.

When the electromagnetic wave is confined between two mirrors, the radiation travels back and forth between the mirrors. We get a field inside the cavity described by a standing wave

$$\mathcal{E}(z,t) = \mathcal{E}_0 \cos \nu t \; \sin kz \,. \tag{4.1.1}$$

The density matrix equations are similar to Eq. (3.5.2-4)

$$\left(\frac{\partial}{\partial t} + v\frac{\partial}{\partial z}\right)\rho_{bb} + \gamma_b \rho_{bb} =$$

$$\lambda_b - i\frac{\mu\mathcal{E}_0}{\hbar} \cos \nu t \; \sin kz\,(\rho_{ba} - \rho_{ab}), \tag{4.1.2}$$

$$\left(\frac{\partial}{\partial t} + v\frac{\partial}{\partial z}\right)\rho_{aa} + \gamma_a \rho_{aa} =$$

$$\lambda_a + i\frac{\mu\mathcal{E}_0}{\hbar} \cos \nu t \; \sin kz\,(\rho_{ba} - \rho_{ab}), \tag{4.1.3}$$

$$\left(\frac{\partial}{\partial t} + v\frac{\partial}{\partial z}\right)\rho_{ba} + (\gamma_{ba} + i\omega)\rho_{ba} =$$

$$- i\frac{\mu\mathcal{E}_0}{\hbar} \cos \nu t \; \sin kz\,(\rho_{bb} - \rho_{aa}). \tag{4.1.4}$$

We assume the transformation

$$\rho_{ba} = \bar{\rho}_{ba}\, e^{-i\nu t}. \tag{4.1.5}$$

We adopt the interaction picture and RWA to get

$$\left(\frac{\partial}{\partial t} + v\frac{\partial}{\partial z}\right)\rho_{bb} =$$

$$(\lambda_b - \gamma_b \rho_{bb}) - i\frac{\mu\mathcal{E}_0}{2\hbar} \sin kz\,(\bar{\rho}_{ba} - \bar{\rho}_{ab}), \tag{4.1.6}$$

$$\left(\frac{\partial}{\partial t} + v\frac{\partial}{\partial z}\right)\rho_{aa} =$$

$$(\lambda_a - \gamma_a \rho_{aa}) + i\frac{\mu\mathcal{E}_0}{2\hbar}\sin kz\,(\bar{\rho}_{ba} - \bar{\rho}_{ab}), \quad (4.1.7)$$

$$\left(\frac{\partial}{\partial t} + v\frac{\partial}{\partial z}\right)\bar{\rho}_{ba} =$$

$$-(i\Delta + \gamma_{ba})\,\bar{\rho}_{ba} - i\frac{\mu\mathcal{E}_0}{2\hbar}\sin kz\,(\rho_{bb} - \rho_{aa}), \quad (4.1.8)$$

where, $\Delta = \omega - \nu$ is the detuning. In the case of a travelling wave, we could treat the space and time dependence together, but in the case of standing wave this is not possible as the atomic position has time dependence,

$$z(t) = z_0 + v\,(t - t_0).$$

The standing wave can be written as two travelling waves propagating in opposite directions.

$$\mathcal{E}(z,t) = \frac{1}{2}\mathcal{E}_0\,[\sin(kz + \nu t) + \sin(kz - \nu t)]. \quad (4.1.9)$$

From physical considerations, we can consider the lowest order effect to be the sum of the effects caused by each wave separately. The induced dipole moment caused by the interaction of the two waves can be written as

$$\rho_{ba} = \rho_+ e^{i(kz - \nu t)} + \rho_- e^{-i(kz+\nu t)}, \quad (4.1.10)$$

such that the two parts ρ_+ and ρ_- are associated with the two oppositely running waves.

Hence, the interaction picture representation is

$$\bar{\rho}_{ba} = \rho_+ e^{i\,kz} + \rho_- e^{-i\,kz}. \quad (4.1.11)$$

The effects of the two terms can be considered separately, and corrections due to the approximation can be considered later. We also assume that the population terms do not have spatial dependence and they can be replaced by their spatial averages.

$$\left(\frac{d}{dt}\rho_+ + ikv\rho_+\right)e^{ikz} + \left(\frac{d}{dt}\rho_- - ikv\rho_-\right)e^{-ikz}$$

$$= -(i\Delta + \gamma_{ba})\,(\rho_+ e^{i\,kz} + \rho_- e^{-i\,kz}) - i\frac{\mu\mathcal{E}_0}{2\hbar}\sin kz\,(\rho_{bb} - \rho_{aa}). \quad (4.1.12)$$

Comparing the coefficients of $e^{i\,kz}$ and e^{-ikz}, we get

$$\frac{d}{dt}\rho_+ = -[i(\Delta + \mathrm{kv}) + \gamma_{ba}]\,\rho_+ - \frac{\mu\mathcal{E}_0}{4\hbar}(\rho_{bb} - \rho_{aa}), \quad (4.1.13)$$

$$\frac{d}{dt}\rho_- = -[i(\Delta - \mathrm{kv}) + \gamma_{ba}]\,\rho_- + \frac{\mu\mathcal{E}_0}{4\hbar}(\rho_{bb} - \rho_{aa}), \quad (4.1.14)$$

$$\frac{d}{dt}\rho_{bb} = (\lambda_b - \gamma_b\rho_{bb}) - i\frac{\mu\mathcal{E}_0}{2\hbar}\sin kz(\rho_+ e^{i\,kz} + \rho_- e^{-i\,kz}$$

$$-\rho_+^* e^{-i\,kz} - \rho_-^* e^{i\,kz})$$

$$= (\lambda_b - \gamma_b\rho_{bb}) + \frac{\mu\mathcal{E}_0}{4\hbar}(\rho_+ + \rho_+^* - \rho_- - \rho_-^*), \quad (4.1.15)$$

$$\frac{d}{dt}\rho_{aa} = (\lambda_a - \gamma_a\rho_{aa}) - \frac{\mu\mathcal{E}_0}{4\hbar}(\rho_+ + \rho_+^* - \rho_- - \rho_-^*). \tag{4.1.16}$$

We have used RWA so that

$$e^{\pm ikz}\sin kz = \pm i/2 .$$

As mentioned earlier we have also dropped the spatial dependence term for the diagonal elements. We assume that the off-diagonal terms reach their equilibrium values much faster the diagonal elements, so that

$$\frac{d}{dt}\rho_\pm = 0.$$

Hence,

$$\rho_\pm = \mp\frac{\mu\mathcal{E}_0}{4\hbar}\frac{(\rho_{bb} - \rho_{aa})}{[\,i(\Delta\pm \mathrm{k}v) + \gamma_{ba}]}. \tag{4.1.17}$$

But

$$(\rho_+ + \rho_+^* - \rho_- - \rho_-^*) = 2Re\,(\rho_+ - \rho_-)$$

$$= \frac{\mu\mathcal{E}_0}{2\hbar}(\rho_{bb} - \rho_{aa})\left[\frac{-\gamma_{ba}}{\gamma_{ba}^2 + (\Delta+kv)^2} - \frac{\gamma_{ba}}{\gamma_{ba}^2 + (\Delta-kv)^2}\right]$$

$$= -\frac{\mu\mathcal{E}_0}{2\hbar\gamma_{ba}}(\rho_{bb} - \rho_{aa})[\,L(\Delta + kv) + L(\Delta - kv)],$$

Where, the dimensionless Lorentzian is

$$L(\Delta \pm kv) = \frac{{\gamma_{ba}}^2}{\gamma_{ba}^2 + (\Delta\pm kv)^2}. \tag{4.1.18}$$

Now assuming steady state of diagonal elements we get

$$\rho_{bb} = \frac{\lambda_b}{\gamma_b} + \frac{\mu\mathcal{E}_0}{4\hbar\,\gamma_b}(\rho_+ + \rho_+^* - \rho_- - \rho_-^*)$$

$$= \frac{\lambda_b}{\gamma_b} + \frac{\mu\mathcal{E}_0}{4\hbar\,\gamma_b}\left(-\frac{\mu\mathcal{E}_0}{2\hbar\gamma_{ba}}\right)(\rho_{bb} - \rho_{aa})[\,L(\Delta + kv) + L(\Delta - kv)]. \tag{4.1.19}$$

Similarly,

$$\rho_{aa} = \frac{\lambda_a}{\gamma_a} - \frac{\mu\mathcal{E}_0}{4\hbar\,\gamma_a}\left(-\frac{\mu\mathcal{E}_0}{2\hbar\gamma_{ba}}\right)(\rho_{bb} - \rho_{aa})[\,L(\Delta + kv) + L(\Delta - kv)]\,, \tag{4.1.20}$$

$$\rho_{bb} - \rho_{aa} = \frac{\lambda_b}{\gamma_b} - \frac{\lambda_a}{\gamma_a} - \frac{1}{2}\left(\frac{\mu\mathcal{E}_0}{2\hbar}\right)^2(\rho_{bb} - \rho_{aa})\frac{1}{\gamma_{ba}}\left(\frac{1}{\gamma_a} + \frac{1}{\gamma_b}\right)$$

$$[\,L(\Delta + kv) + L(\Delta - kv)], \tag{4.1.21}$$

or,

$$\rho_{bb} - \rho_{aa} = \frac{\bar{N}}{1 + I\eta/2[\,L(\Delta+kv) + L(\Delta-kv)]}, \tag{4.1.22}$$

where,

$$\bar{N} = \frac{\lambda_b}{\gamma_b} - \frac{\lambda_a}{\gamma_a}, \tag{4.1.23}$$

$$I = \frac{1}{2\,\gamma_b\,\gamma_a}\left(\frac{\mu\mathcal{E}_0}{\hbar}\right)^2, \tag{4.1.24}$$

$$\eta = \frac{\gamma_a + \gamma_b}{2\gamma_{ba}}. \tag{4.1.25}$$

As discussed in the case of interaction with a progressive wave, the **unsaturated population difference** $\bar{N}$ is described by a Gaussian distribution function; the dimensionless field intensity and the saturation parameters are the same as described in section 3.5. Hence, the population difference, given by Eq. (4.1.22), is a Gaussian function with two holes described by the Lorentzian function centred at $v = \pm(\omega - \nu)/k$ and with a width of γ_{ba}. This is described in Fig. 4.1. A comparison with the interaction of a progressive wave discussed in Chapter 3 (Fig. 3.4) shows that the two holes arise from two oppositely moving waves. They are away from the centre by a magnitude determined by the amount of detuning. The non-diagonal matrix elements represented by $\rho_\pm$ in Eq. (4.1.17) are

$$\rho_\pm = \mp\frac{\mu\mathcal{E}_0}{4\hbar}\frac{\bar{N}\,[-\,i(\Delta\pm kv)+\gamma_{ba}]}{[1+I\eta/2[\,L(\Delta+kv)+L(\Delta-kv)][\gamma_{ba}^2+(\Delta\pm kv)^2\,]}. \tag{4.1.26}$$

From Eq. 4.1.10 and 4.1.26 we can write

$$\rho_{ba}(z,t) = \frac{\mu\mathcal{E}_0}{4\hbar}\frac{\bar{N}e^{-i\nu t}}{[1+I\eta/2[\,L(\Delta+kv)+L(\Delta-kv)]}\Big[\frac{[\,i(\Delta+kv)-\gamma_{ba}]e^{ikz}}{[\gamma_{ba}^2+(\Delta+kv)^2]}$$

$$+\frac{[-\,i(\Delta-kv)+\gamma_{ba}]e^{-ikz}}{[\gamma_{ba}^2+(\Delta-kv)^2]}\Big]. \tag{4.1.27}$$

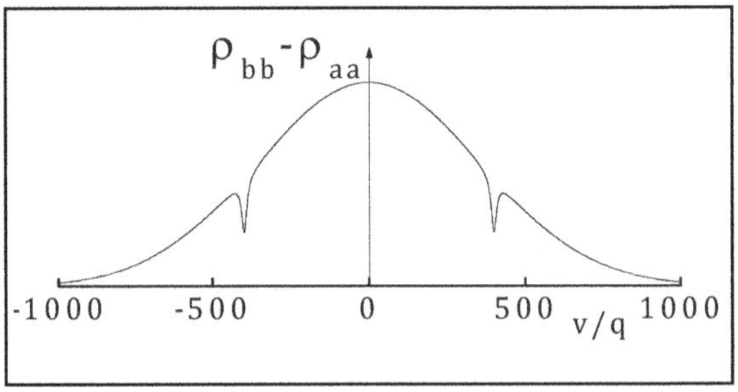

Fig. 4.1 The holes created by the oppositely running waves in the velocity distribution curve.

It is evident from Fig 4.1 for population difference that the standing wave consisting of two oppositely running travelling waves burns two independent population holes (Bennett hole, see chapter 3) in the velocity distribution curves. The holes are situated at $v = \pm\Delta/k$ and their heights are determined by the field intensity I and the saturation parameter η. With increase of detuning, the holes move away from each other until they reach the tail of the velocity distribution curves. This is based on our assumption that the two waves are independent of each other such that they do not interact. When the detuning is small, they approach each other and the two holes may overlap at resonance. In such a case, we may have to consider the interacting waves. But in our approximation, physics of the problem is still correctly described.

4.2 Lamb dip

We noted in the last section that if the detuning is reduced the two Bennett holes burnt on the velocity distribution curve approach each other. When the detuning eventually goes to zero the holes coincide and a strong dip appears (Fig. 4.2). This is called **Lamb dip**. Physical process of formation of Lamb dip is as follows. One of the beams depletes $N_+(v)$ atoms of the velocity class around the velocity $v_+ = +\Delta/k$. At the same time, the other beam interacts with a group of atoms around the velocity class $v_- = -\Delta/k$. When the velocity classes are different these atoms are distinct, the presence of one beam does not affect the absorption of the other beam. But if $\Delta = 0$, we have $v_+ = v_-$, in this case, the two beams interact with all the atoms of same zero velocity class.

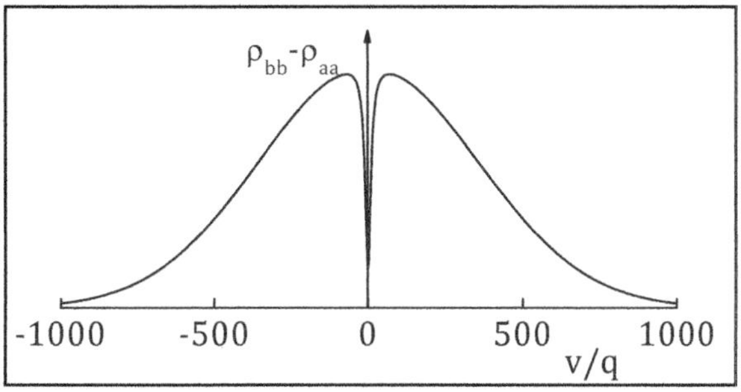

Fig. 4,2 The Lamb dip at the central frequency where the two holes merge together at resonance.

It is clear from Eq. (4.1.22) that close to the resonance frequency, i.e. when $\Delta \rightarrow 0$, the two Lorentzian functions become the same for all velocity classes. One beam will meet the depletion of the atoms caused by the other beam, called the pump beam. This leads to large reduction of absorption intensity as the ground state will have much less population at the resonance frequency.

The dip has width comparable to the natural width of the transition and can be used for measurement of hyperfine transition frequency. If there are closely spaced transitions with frequency spacing much smaller than the Doppler width, they cannot be resolved in a single wave spectroscopy. But they can be easily resolved in the standing wave spectroscopy. We shall describe in Sec. 4.6 an experimental arrangement for observation of Lamb dip in Rubidium atom and demonstrate how the hyperfine structure can be resolved.

4.3 Crossover resonance dip

In the case of an atom with close lying transitions at frequencies $\omega_1$and ω_2, the standing wave will produce two sets of holes (Fig. 4.3) at velocities $\pm\Delta_1/k$ and $\pm\Delta_2/k$, where $\Delta_i = (\omega_i - \nu)$ (i=1,2) for each of the transitions. The positions of the holes will depend on the laser frequency.

When the laser frequency ν is tuned, it will approach the frequency ω_1; the two dips will coincide and we have a Lamb dip as is discussed earlier. Similar will be the case of ω_2 in course of tuning the laser frequency. However, in the process of tuning, one dip on the left hand side at velocity $-\Delta_1/k$ may coincide with the dip at velocity at $+\Delta_2/k$. Similarly, the two dips at the velocities at $-\Delta_2/k$ and $+\Delta_1/k$ may coincide in the course of tuning the laser radiation. Such coincidences may occur at $\Delta_1 = -\Delta_2$. This will lead to a new resonance at $\nu = (\omega_1 + \omega_2)/2$.

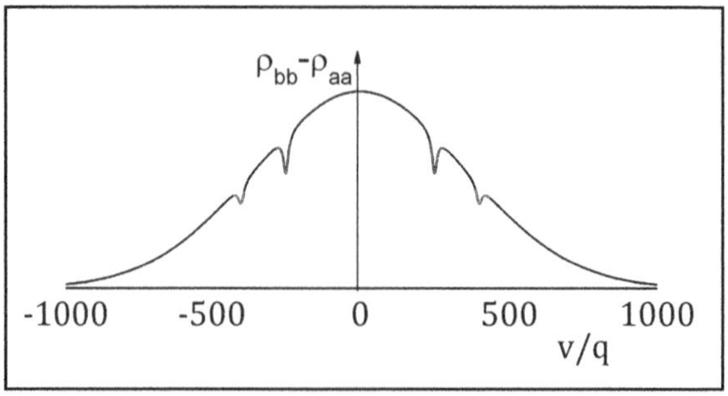

Fig. 4.3 Population difference holes created by the laser due to two transitions, marked red and green.

This phenomenon can be explained, if we consider the velocity class of atoms, such that $\nu + kv = \omega_2$ and $\nu - kv = \omega_1$. One beam excites the atoms from the ground state and depletes them from the ground state, while the other beam will find less atoms of the same velocity in the ground state and hence will have a reduced absorption. Similarly, the atoms with velocities satisfying the relation $\nu + kv = \omega_1$ and $\nu - kv = \omega_2$ will also have a resonance at $\nu = (\omega_1 + \omega_2)/2$. This dip, called the **Crossover transition dip**, can be explained by the energy level diagram as shown (Fig. 4.4). Thus, in an atomic system with two excited close-lying states and a ground state, there will be three dips corresponding to the two transitions and a crossover dip (Fig. 4.5). In a system with n number of excited states, there will be n Lamb dips and $n(n-1)/2$ crossover transition dips. In case of three upper levels, there will be three crossover resonance dips in addition to the three Lamb dips.

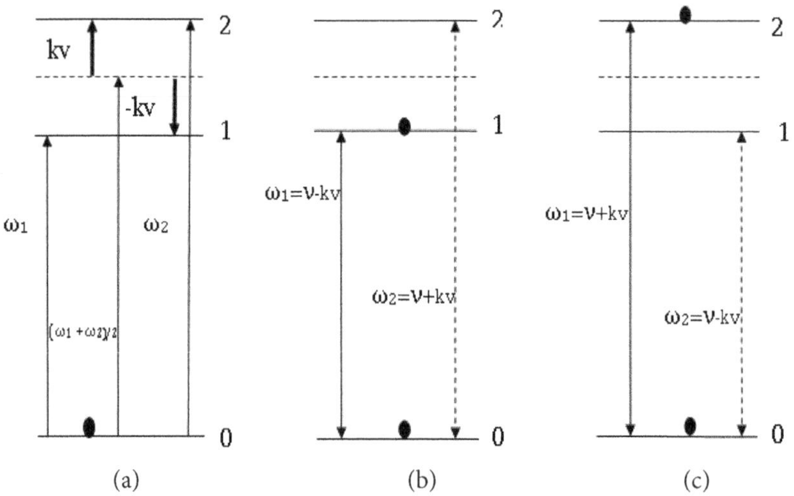

Fig. 4.4 Energy level diagram showing the hyperfine and crossover transition for two closely spaced upper levels, (a) presents two hyperfine and one crossover transitions. Figures (b) and (c) show how the holes for the two counter propagating waves arising from two different transitions can meet, when $\nu = (\omega_1 + \omega_2)/2$.

4.4 Closed transitions and Optical Pumping

We consider a four level system with two upper levels 3 and 4 and two lower levels 1 and 2 (Fig. 4.6). The spacing between the lower levels is such that a single laser beam cannot excite atoms from both of these states at a time. The transitions $1 \rightarrow 3$, $2 \rightarrow 3$ and $2 \rightarrow 4$ are allowed by selection rule, while the transition from the ground state 1 to the level 4 is not allowed.

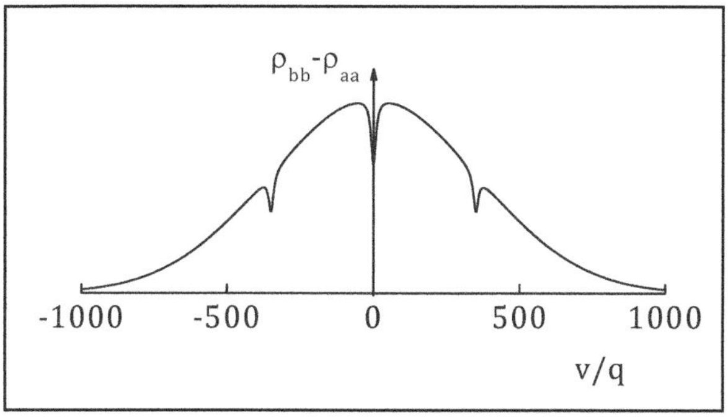

Fig. 4.5 Two hyperfine transitions with a central crossover transition.

If we consider all the atoms to be initially in the ground state 2, a laser with the frequency corresponding to the 2→4 transition will excite some atoms to the level 4. The atoms from this state can decay back to the level 2, but they cannot decay to the level 1 because of the selection rule. Hence the atoms remain confined to the states 2 and 4 and we call the transition a **Closed Transition.**

If we consider again the atoms in the ground state 2 and a laser with appropriate frequency is tuned to the transition 2→ 3, then some of the atoms will be excited to the level 3. The atoms from this state can decay to both the levels 1 and 2 because of the selection rule. The atoms that are transferred to level 1 will remain confined to this level, since the laser is tuned at a frequency far from the allowed transition frequency from the level 1 to the level 3.

Thus, there is no way that some of these atoms can be transferred back to the level 2. Hence, after a few absorption-spontaneous emission cycles, the atoms will eventually be transferred to the ground state 1. The system will become transparent. This phenomenon is known as **Optical Pumping.** It is a process in which atoms are transferred from one ground level to another by means of optical absorption from the first level.

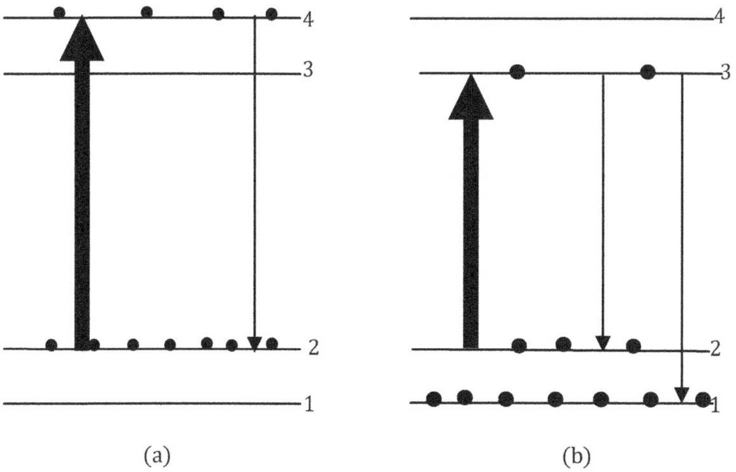

Fig. 4.6 (a) Closed transition and (b) Optical pumping

4.5 Atomic energy levels of Rubidium

Rubidium is an alkali metal with two isotopes Rb^{85} and Rb^{87}, with natural abundances of 72.2% and 27.8% respectively. They have nuclear spins I = 5/2 for Rb^{85} and I= 3/2 for Rb^{87}. The ground state electronic configuration is the completely filled core of inert gas atom Kr surrounded by a single valence electron in the 5s state with the electron configuration

1s(2)2s(2)2p(6)3s(2)3p(6)3d(10)4s(2)4p(6)5s.

The numbers in the bracket indicate the number electrons in the sub-shell. The next higher states correspond to the higher values of orbital angular momentum quantum number like 5p, 5d. 5f etc. for the same Rydberg state n=5. The theory of angular momentum and their interaction have been treated in details in standard text-books on Quantum Mechanics. We discuss angular momenta and their interaction with special reference to Rb.

Fine Structure

The lone electron moving in the outermost orbit faces an electric field produced by the combined effect of the nuclear charge and the electric charge of the core electrons. The electron has an orbital angular momentum $\hat{\boldsymbol{L}}$ with the components $\hat{L}_x, \hat{L}_y, \hat{L}_z$. The uncertainty principle of quantum mechanics does not allow commutation of the components and hence all three components cannot be measured. It can be shown that only $\hat{L}^2$ and one component of $\hat{L}$ can be simultaneously measured, since they commute. The measured eigenvalues are $l(l+1)\hbar^2$ and $m_l\hbar$. The orbital angular momentum l can have values 0,1,2 etc corresponding to s, p, d, states, m_l can have $2l$ +1 integral values from $-l$ to $+l$.

The electron has an intrinsic spin angular momentum and it can be described in a similar way with the square of the spin angular momentum $\hat{S}^2$ and one component $\hat{S}_z$. The spin for electron is $S = ½$. The measured eigenvalues are $S(S+1)\hbar^2 = \frac{3}{4}\hbar^2$ and $m_s\hbar$ where m_s can have values +½ and -½. The magnetic moments corresponding to the orbital and spin angular momenta interact with each other and this leads to a splitting of the energy levels.

For an atom with a large number of electrons, we can add the angular momenta. The total orbital angular momentum of an atom containing many electrons is the sum of the orbital angular momentum of each electron

$$\boldsymbol{L} = \sum_i \boldsymbol{l}_i. \tag{4.5.1}$$

Similarly the total spin angular momentum is given by the sum of the spin angular momentum of each electron as

$$\boldsymbol{S} = \sum_i \boldsymbol{s}_i. \tag{4.5.2}$$

The total electronic angular momentum of the atom is then

$$\boldsymbol{J} = \boldsymbol{L} + \boldsymbol{S}. \tag{4.5.3}$$

An important result is that the total angular momentum of a sub-shell with completely filled electrons is zero. In this case, for each electron with a non-zero value $+m_l$, there is an electron with a value $-m_e$; similarly, for each electron with a value $+m_s$ there is an electron with a value $-m_s$. Thus, the total electronic angular momentum component $J_z = 0$. Since the only allowed value of J_z is zero, the total angular momentum of the completely filled core is zero. For rubidium atom with a single unpaired electron in the valence shell, the orbital angular momentum is L and the spin angular momentum is ½ . The possible values of the total angular momentum quantum number are $J = |L \pm 1/2|$.

Different values of J arise from different relative orientation of the vectors $\vec{L}$ and $\vec{S}$. There is an interaction of the magnetic dipole moment associated with the orbital angular momentum and the intrinsic spin angular momentum. This leads to a term proportional to $\hat{\boldsymbol{L}} \cdot \hat{\boldsymbol{S}}$ in the interaction energy. Since

$$\hat{\boldsymbol{L}} \cdot \hat{\boldsymbol{S}} = \frac{\hat{J}^2 - \hat{L}^2 - \hat{S}^2}{2}, \tag{4.5.4}$$

$$\Delta E_{ls} \propto \frac{1}{2}[J(J+1) - L(L+1) - S(S+1)]. \tag{4.5.5}$$

The relative values of the fine structure shift may thus be computed for the state with a set of L, S, J values.

In the case of Rb atom, the ground state is 5s. Since $L = 0$, there is no spin-obit interaction, so there is no splitting. The next higher state 5p has $L = 1$, hence the values of J are 3/2 and 1/2. The **fine structure splitting** between the two successive values of J is the difference between the energies of the levels with J and J-1. Hence,

$$\Delta E\,(J) - \Delta E\,(J-1) \infty J(J+1) - (J-1)J = 2J. \tag{4.5.6}$$

Thus, the fine structure splitting is proportional to the total angular momentum of the upper level. This is known as the Lande interval rule. In the case of Rb atom, the 5s-5p transition shows splitting into D_1 ($^2S_{1/2}$ -$^2P_{1/2}$) and D_2 ($^2S_{1/2}$ -$^2P_{3/2}$) transitions (Fig. 4.7), a phenomenon that is true for all alkali atoms. In the case of Rb, this splitting corresponds to a wavelength difference of 15 nm. Hence, in a single scan of the laser frequency with a diode laser, both the transitions cannot be observed by scanning one laser frequency. The usual diode laser can be tuned over a much shorter wavelength range.

Hyperfine structure

We have, so far, considered the fine structure splitting caused by the spin-orbit interaction. We have also to consider the interaction between the magnetic moments associated with the total angular momentum and the nuclear magnetic moment. This interaction leads to the coupled angular momentum as

$$\hat{\boldsymbol{F}} = \hat{\boldsymbol{J}} + \hat{\boldsymbol{I}} \tag{4.5.7}$$

and the quantum number F can have values from $|J - I|$ to J+I, varying in integral steps.

In the case of Rb^{87}, $I = 3/2$. For the ground state $J = ½$, hence the state is split into two levels with F = 1,2. For the excited state with $J = 3/2$, there is a splitting into for levels F = 0, 1, 2, 3. Similarly, for the P-state with J = ½, the hyperfine splitting is into two levels with F = 1,2. The energy of the hyperfine level can be expressed in terms of the energy of the unsplit J-th level by the Casimir formula involving two hyperfine constants A and B

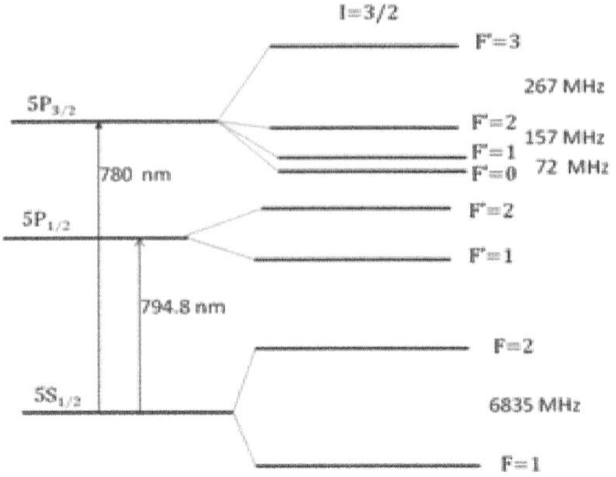

Fig. 4.7 Energy level diagram showing D_1 and D_2 transitions of Rb^{87}. The energy levels are not to scale.

$$E_F = E_J + A\frac{\kappa}{2} + B\frac{3\kappa(\kappa+1)/4 - I(I+1)J(J+1)}{2I(2I-1)J(2J-1)} \tag{4.5.8}$$

$$\kappa = F(F+1) - J(J+1) - I(I+1) \tag{4.5.9}$$

If $I = 1/2$ or $J = 1/2$, the third term containing B is not to be included.

The **hyperfine** energy level splitting is shown in Fig. 4.8. The separation between the levels also follows the Lande interval rule and it is proportional to the F value of the upper level. In the case of Rb^{85}, the lower state $S_{1/2}$ will split into levels F = 2,3 and the two upper states will split into the states with F = 2,3 and F = 1,2,3,4.

Exercise 4.1 Calculate the hyperfine splitting of energy levels of Rb^{85}.

(A= 1011.91 *and* 120.72 MHZ *for* $S_{1/2}$ *and* $P_{1/2}$ levels respectively and $A = 25.01$ and $B = 25.88\ MHz$ for $P_{3/2}$ level.).

Solution

For Rb^{85} , I=5/2, $S_{1/2}$ level, A= 1011.91 MHz

$P_{1/2}$ level, A= 120.72 MHz

$P_{3/2}$ level, A= 25.01 MHZ

B = 25.88 MHz

From Eq. (4.5.8) and (4.5.9), we get energy for the hyperfine components as

$$E_F = E_J + A\frac{\kappa}{2} + B[\frac{3\kappa(\kappa+1)}{8I(2I-1)J(2J-1)} - \frac{I(I+1)J(J+1)}{2I(2I-1)J(2J-1)}]$$

where, $\kappa = F(F+1) - J(J+1) - I(I+1)$.

For $S_{1/2}$ state, F=3, 2 for F=3, κ=5/2, for F=2, $\kappa = -7/2$.

It can easily be shown that

$$E_3 - E_2 = 1011.91\left[\frac{5}{2}+\frac{7}{2}\right] = 3035.73\ MHz$$

Similarly, for $P_{1/2}$, state, F=3, 2 , we can show that

$$E_3 - E_2 = 120.72\left[\frac{5}{2}+\frac{7}{2}\right] = 362.16\ MHz$$

For $P_{3/2}$ state, F= 4,3,2,1. For each value of F we can calculate κ and hence calculate the hyperfine energy differences with the given values of A and B.

In laser absorption spectroscopy, we can see a set of transitions from either of the lower levels to the three upper levels following the selection rules $\Delta F = 0, \pm 1$. The transitions D_1 and D_2 are spectrally separated. The three transitions starting from one particular lower state to the three upper hyperfine states will be separated by the hyperfine splitting of the upper levels. The splitting is much smaller than the Doppler broadening observed in a room temperature experiment performed by scanning a single laser. So the transition frequencies will all be within the Doppler background and hence cannot be resolved. With one sample of Rb we would observe four Doppler broadened transitions. Two such unresolved transitions from the lower hyperfine states for each isotope are observed. The hyperfine structure splitting for the ground state (Fig. 4.8) being larger than the Doppler broadening, all the four transitions are distinctly separated (Fig.4.9).

4.6 Saturation Spectroscopy Experiment

Lamb dip or saturation absorption spectroscopy experiments on Rb are carried out with low power frequency tunable diode lasers. The laser frequency depends on laser temperature and laser current. The current and temperature can be set for providing coarse tuning over a range of wavelength. In high resolution

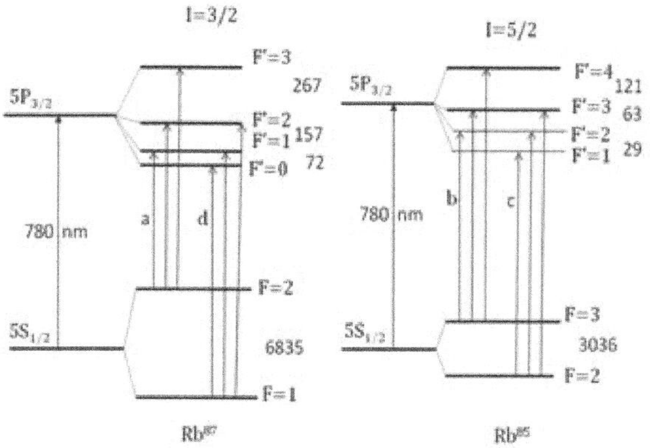

Fig. 4.8 Energy level splitting and hyperfine transitions for the D_2 transition in both isotopes of Rb. All energy differences shown are in MHz unit.

spectroscopy experiments, **external cavity diode lasers** (ECDL) are used (Fig. 4.10). In ECDL, the laser radiation is reflected by a grating to a tuning mirror that can be controlled by a piezoelectric transducer. The grating and the mirror together form an external cavity. The number of half wavelengths that can be sustained within the cavity is determined by the cavity length. The PZT can be tuned externally by a computer controlled PZT transducer voltage (0-150V). The PZT pushes the mirror; as a result, the cavity length changes and consequently the wavelength changes.

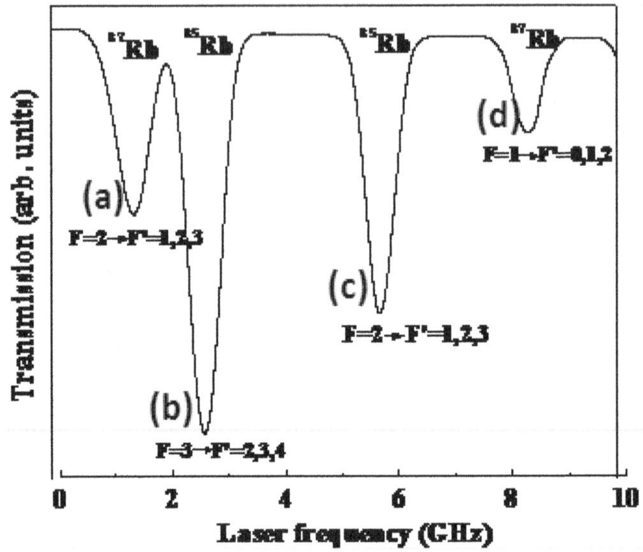

Fig. 4.9 Doppler broadened transitions for two isotopes of Rubidium as shown in Fig. 4.8.

Thus, the tuning of the mirror provides fine tuning of the laser frequency. If the mirror is tuned too far away the cavity length change is high and the number of half-wavelengths jumps up or down. This will cause the laser to set back to the original range of frequency set by the temperature and current of the diode. This leads to a mode hop. A new mode starts with a a new arbitrary phase. Such mode breaks were discussed in Chapter 1, while discussing coherence properties of laser radiation.

The coarse tuning of the laser is provided by the temperature and current control. By a careful choice of temperature and current the grating can be tuned over the range of nearly 8 GHz, so that the four Doppler broadened profiles for the hyperfine group of transitions for the two isotopes can be observed without a mode hop (Fig. 4.9).

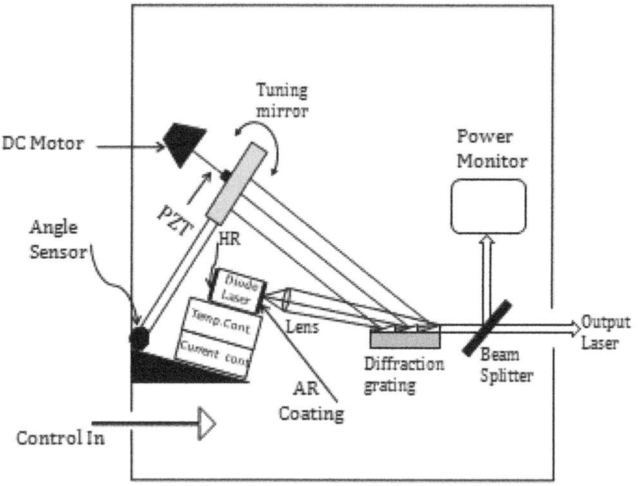

Fig. 4.10 External Cavity Diode Laser (ECDL) setup

In the case of saturation absorption spectroscopy, as shown in Fig 4.11, the laser radiation from the external cavity diode laser is split into two fractions by a beam splitter. The reflected part is sent to the Rb cell and constitutes the forward moving probe beam. The other fraction is transmitted through the beam splitter. It is reflected by two mirrors and the reflected wave passes through the Rb cell as a counter-propagating beam, called the pump beam. The beam is sent to a beam dump that stops the beam. The probe is recorded by a photo-detector and exhibits the effect of pump probe saturation spectroscopy. Another beam (not shown in the figure) reflected from the back surface of the beam splitter also passes through the cell as a probe beam that does not meet the pump beam. The probe beams are sent to the balanced photo-detector. The spectra recorded show the Lamb dips and crossover resonance dips.

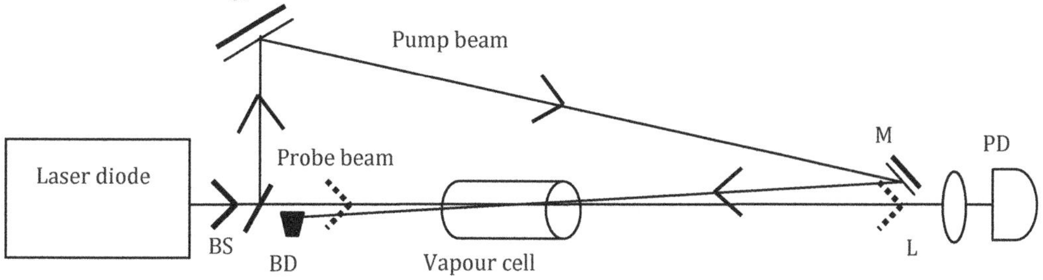

Fig. 4.11 Block diagram for recoding the Lamb dip and crossover resonance dips by Saturation Absorption Spectroscopy. BS – Beam Splitter, M – Mirror, BD – Beam Dump, PD – Photo Detector, L – Lens.

Fig. 4.12 shows the observed transitions for Rb^{85} isotope. For the set of three hyperfine transitions, it is expected that one can record three more transitions depicting the crossover transitions as mentioned in Sec 4.3. However the hyperfine splitting for the isotope is small.

So they overlap as shown in the energy level diagram. This is more evident for the lower hyperfine levels of the

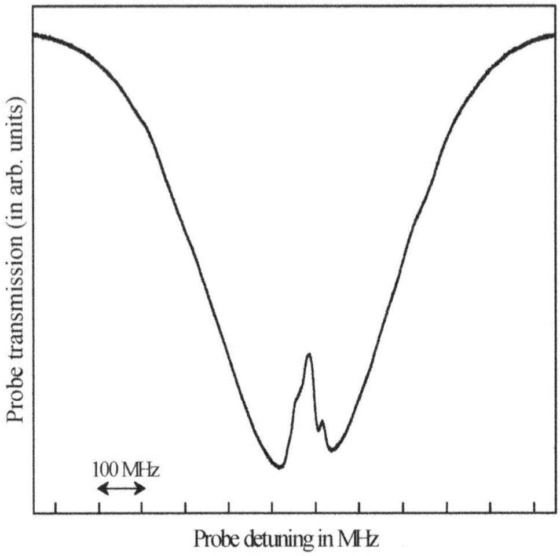

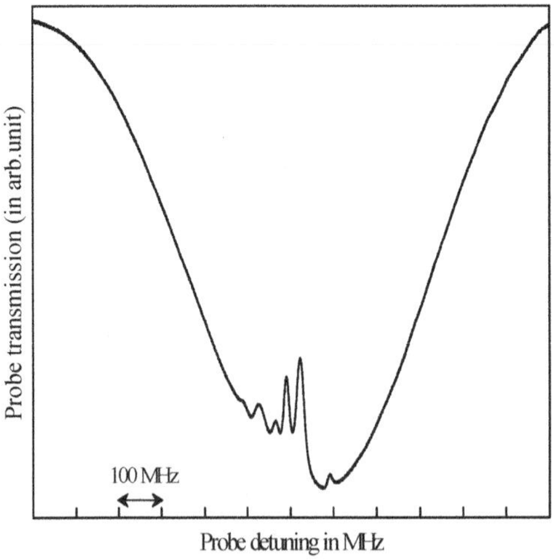

(F=2 → F'=1, 2, 3) (F=3 → F'=2, 3, 4)

Fig. 4.12 The saturation spectra of Rb^{85} isotope.

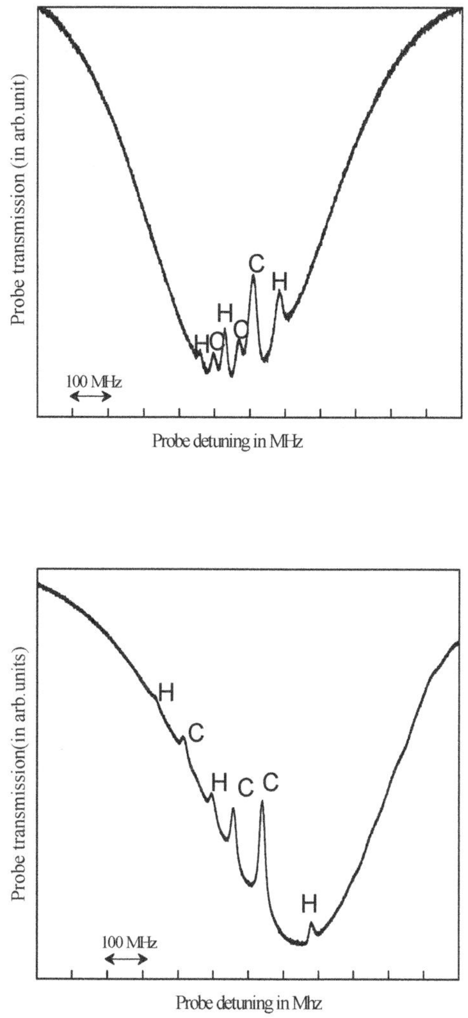

(F=1→ F'=0, 1, 2) (F=2 → F'=1, 2, 3)

Fig. 4.13 The saturation spectra of Rb^{87} isotope.

F=2 → F'=1, 2, 3 transition. For the other hyperfine group (beginning F=3), they are partially resolved. Hence they are not identified in the spectrum. For the Rb^{87} transitions, the hyperfine splitting is higher and all the six transitions, three hyperfine (marked as H) transitions and the three crossover resonances (marked as C) are fully resolved. However, in diode laser spectroscopy usually etalon fringes are used for calibration.

Problems

4.1 Solve the rate equation for $\rho_{\pm}$ in the case of $\Delta = \mp kv$. Obtain expression for the peak heights of the absorption at these frequencies in terms of Laser intensity and decay constants.

4.2 Show that the peak heights of the two holes at $\Delta = \mp kv$ are the same. What happens to the two holes when $\Delta = 0$?

4.3 The rate equations for the density matrix of a two-level stationary atom in a standing wave $\mathcal{E}(z,t) = \mathcal{E}_0 \cos \nu t \sin kz$ are

$$\frac{d\rho_{bb}}{dt} + \gamma_b \rho_{bb} = \lambda_b - i\frac{\mu\mathcal{E}_0}{\hbar} \cos \nu t \sin kz\, (\rho_{ba} - \rho_{ab}),$$

$$\frac{d\rho_{aa}}{dt} + \gamma_a \rho_{aa} = \lambda_a + i\frac{\mu\mathcal{E}_0}{\hbar} \cos \nu t \sin kz\, (\rho_{ba} - \rho_{ab}),$$

$$\frac{d\rho_{ba}}{dt} + (\gamma_{ba} + i\omega)\rho_{ba} = -i\frac{\mu\mathcal{E}_0}{\hbar} \cos \nu t \sin kz\, (\rho_{bb} - \rho_{aa}).$$

Using RWA rewrite the equations. State the conditions under which perturbative time dependent solutions are possible.

4.4 Solve the equations of problem 4.3 under steady state and show that the population difference is

$$\rho_{bb} - \rho_{aa} = \frac{\bar{N}}{1 + 2I\eta L(\Delta)\sin^2 kz}$$

The constants are as defined in the text. $L(\Delta)$ is a dimensionless Lorentzian as defined above. Show that the deviation from $\bar{N}$ is a Lorentzian as a function of detuning with width

$$\gamma_{eff} = \gamma_{ba}[1 + 2I\eta \sin^2 kz\,]^{1/2}.$$

Give plots of the sinusoidal functions for ρ_{bb} and ρ_{aa} against z.

4.5 Solve the rate equations of problem 4.3 assuming $\rho_{ba} = \bar{\rho}_{ba}\, e^{-i\nu t}$ and rapid relaxation to the steady state $\dot{\bar{\rho}}_{ba} = 0$ show that the non-diagonal matrix element can be written as

$$\bar{\rho}_{ba} = -\frac{\bar{N}\Omega \sin kz(\Delta + i\gamma_{ba})}{2[\Delta^2 + \gamma_{eff}{}^2]}.$$

4.6 Using the results of problem 4.3-5 obtain an expression for polarization in terms of laser intensity.

4.7 Draw an energy level diagram with fine and hyperfine splitting to show the Rb D_1 transitions. For one isotope how many hyperfine Lamb dips for hyperfine transitions are possible? What are the crossover resonances in this case?

4.8 Draw an energy level diagram to show the fine and hyperfine structure components for Cs. Mark the quantum numbers and show the Lamb dip and crossover resonance dips in saturated absorption spectroscopy.

4.9 The energies of the hyperfine levels relative to the fine structure energy levels can be expressed by the Casimir formula in terms of two constants A and B as given in the text.

Calculate the hyperfine splitting of energy levels for Rb^{87}.

(A= 3417.34 and 406.20 $MHZ\ for\ S_{1/2}\ and\ P_{1/2}$ levels respectively and $A = 84.80$ and $B = 12.52\ MHz$ for $P_{3/2}$ level.).

4.10 If the energy difference between 5s and 5p levels of Rb is 1.6 eV, find an estimate for the fine structure and hyperfine structure level splitting in eV.

Further reading

1. S.C.Rand, Non-linear and Quantum Optics, Oxford, New York, 2010.
2. S. Stenholm, Foundations of Laser Spectroscopy, Wiley-Interscience, New York, 1983.
3. L. Allen and J. H. Eberly, Optical Resonance and Two-level Atoms, Dover, New York, 1987.
4. E. Condon and G. Shortley, The Theoty of Atomic Spectra, Cambridge University Press, Cambridge, 1951.
5. C. Cohen-Tanoudji, B. Diu and F. Laloe, Quantum Mechanics Vol I and II, Wiley-Interscience, NY, 1977.
6. V. S. Letokhov and V. P. Chebotayev, Non-linear Laser Spectroscopy, Springer, New York, 1977.
7. W. Demtroeder, Laser Spectroscopy, Springer, Heidelberg, 1981.

Chapter 5
Semi classical Theory of Laser Action

5.1 Self-consistent theory of laser

Basic physical principle of laser action was introduced in Chapter 1. We interpreted the laser action as a result of amplification by multiplying the number of photons inside the cavity containing the medium (atoms) and feedback of the energy to build up sustained oscillation. In this chapter we attempt a semi classical model in terms of a self consistent theory. An electric field present within the medium produces microscopic dipole moments; these moments added together in a many-atom system lead to a macroscopic polarization. This polarization acts as a source in the Maxwell's equations producing a reaction field. We demand that the reaction field is the same as the field that induced the process. This is the condition of self-consistency. The process can be described by a flow chart (Fig. 5.1) for the semi classical model. The electric field produces a dipole moment that can be explained by using the quantum mechanical model of the interaction of the classical electromagnetic field with an atom having quantized energy levels. Principles of statistical mechanics are used by considering the medium as an ensemble and density matrix is used to facilitate the process of calculating the macroscopic polarization.

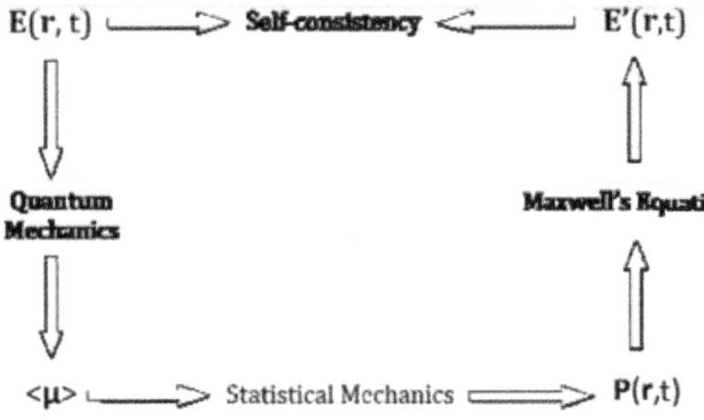

Fig. 5.1 Self-consistent model where the generating field E(r,t) is equal to the reaction field E'(r,t).

We present the semi classical theory developed by Lamb. We shall treat a single longitudinal mode of a scalar plane wave electromagnetic field and a homogeneously broadened medium containing two-level atoms. This theory can include decay phenomena. It can explain the intensity and the frequency of the laser oscillation. It can apply to the lasers bounded by Brewster windows in a cavity that can sustain oscillation of a linearly polarized field. We assume the electric field to be plane-polarized and we ignore the variation of the field transverse to the laser axis. Inside the cavity the electromagnetic field builds up till it can saturate the gain down to the cavity losses.

5.2 Maxwell's field equations

The electromagnetic field inside the cavity is governed by the Maxwell's equations.

$$\nabla \times \mathcal{H} = J + \frac{\partial D}{\partial t}, \tag{5.2.1}$$

$$\nabla \times \mathcal{E} = -\frac{\partial B}{\partial t}, \tag{5.2.2}$$

$$\nabla . \boldsymbol{D} = 0, \tag{5.2.3}$$

$$\nabla . \boldsymbol{B} = 0, \tag{5.2.4}$$

where,

$$\boldsymbol{D} = \epsilon_0 \mathcal{E} + \boldsymbol{P}, \tag{5.2.5}$$

$$\boldsymbol{B} = \mu_0 \mathcal{H}, \tag{5.2.6}$$

$$\boldsymbol{J} = \sigma \mathcal{E}. \tag{5.2.7}$$

We have assumed the presence of a lossy medium with conductivity σ that accounts for diffraction and transmission through the reflector. As in standard notation, ϵ_0 and μ_0 are permittivity and permeability of the medium that can have a polarization $\boldsymbol{P}$. From the above set of equations (5.2.1-7), we can write

$$\begin{aligned} \nabla \times \nabla \times \mathcal{E} &= -\mu_0 \frac{\partial}{\partial t}(\nabla \times \mathcal{H}) \\ &= -\mu_0 \frac{\partial}{\partial t}\left(\boldsymbol{J} + \frac{\partial \boldsymbol{D}}{\partial t}\right) \\ &= -\mu_0 \sigma \frac{\partial \mathcal{E}}{\partial t} - \mu_0 \epsilon_0 \frac{\partial^2 \mathcal{E}}{\partial t^2} - \mu_0 \frac{\partial^2 \boldsymbol{P}}{\partial t^2}. \end{aligned} \tag{5.2.8}$$

Again, we have

$$\nabla \times \nabla \times \mathcal{E} = -\nabla^2 \mathcal{E} + \nabla (\nabla . \mathcal{E}) = -\nabla^2 \mathcal{E}. \tag{5.2.9}$$

We have assumed $\nabla . P = 0$, since P is a lower order term and from Maxwell equation $\nabla . D = 0$. The field intensity varies slowly in the transverse direction, so we consider variation in the z-direction only. Hence

$$\frac{\partial^2 \mathcal{E}}{\partial z^2} = \mu_0 \sigma \frac{\partial \mathcal{E}}{\partial t} + \mu_0 \epsilon_0 \frac{\partial^2 \mathcal{E}}{\partial t^2} + \mu_0 \frac{\partial^2 \boldsymbol{P}}{\partial t^2}. \tag{5.2.10}$$

5.3 Expansion in normal modes

The electric field inside the cavity can be expanded in terms of normal modes. We shall discuss in Chapter 6 that only certain discrete modes achieve appreciable magnitude within the cavity. If the cavity length is *L,* the cavity frequencies are

$$\Omega_n = \frac{n\pi c}{L} = K_n c, \tag{5.3.1}$$

n is a large integer typically of the order of 10^6 and c is velocity of light, K_n is the corresponding wavenumber. If the length L is large, the distance between the successive mode frequencies becomes small and in the case of a cavity with no walls, we have a continuum of modes. Eq. (5.2.8) suggests that the time and space dependent parts can be separated. In terms of normal modes the electric field can be written as

$$\boldsymbol{\mathcal{E}}(z,t) = \frac{1}{2}\sum_n \boldsymbol{\Lambda}_n(t) U_n(z) + c.c., \tag{5.3.2}$$

$\boldsymbol{\Lambda}_n$ is the complex electric field vector for the mode n. We can set

$$\boldsymbol{\Lambda}_n(t) = \mathcal{E}_n(t) \exp\{-i(\nu_n t + \phi_n(t)), \tag{5.3.3}$$

where the electric field amplitude $\mathcal{E}_n(t)$ and the phase $\phi_n(t)$ vary slowly in an optical time period, $\nu_n + \dot{\phi}_n$ is the oscillation frequency of the mode. We shall show that this frequency is not necessarily the same as the cavity frequency when the cavity is filled with a medium. The normal modes can be represented by a sinusoidal z-dependence like $U_n(z) = A\sin K_n z$, in the case of a standing wave with two reflecting mirrors. It can be represented as exp $(iK_n z)$ for a unidirectional laser, e.g. a ring laser. Our expansion of the electric field in the z direction is

$$\mathcal{E}(z,t) = \frac{1}{2}\sum_n \mathcal{E}_n(t)\exp\{-i(\nu_n t + \phi_n)\}U_n(z) + c.c. \tag{5.3.4}$$

$\mathcal{E}_n(t)$ is now the amplitude of the electric field component in the z direction. So it is not a vector. The induced polarization of the medium can be written as

$$P(z,t) = \frac{1}{2}\sum_n P_n(t)\exp\{-i(\nu_n t + \phi_n)\}U_n(z) + c.c. \tag{5.3.5}$$

$P_n(t)$ is a complex slowly varying function of time and represents the polarization for the mode n. The real part of polarization is in phase with the electric field and results in dispersion, while the imaginary part is orthogonal and leads to gain or loss.

5.4 Lamb's self-consistency equations

Substituting the expansions of $\mathcal{E}(z,t)$ and $P(z,t)$ without the complex conjugates into the wave equation (Eq. 5.2.10) we get

$$-K_n^2\mathcal{E}_n(t) = \mu_0\sigma[\dot{\mathcal{E}}_n - i(\nu_n + \dot{\phi}_n)\mathcal{E}_n] +$$

$$\mu_0\epsilon_0[\ddot{\mathcal{E}}_n - 2i\left((\nu_n + \dot{\phi}_n)\dot{\mathcal{E}}_n - i\ddot{\phi}_n\mathcal{E}_n - (\nu_n + \dot{\phi}_n)^2\mathcal{E}_n\right] + \mu_0[-(\nu_n + \dot{\phi}_n)^2 P_n - i(\nu_n + \dot{\phi}_n)\dot{P}_n]. \tag{5.4.1}$$

In the case of a dilute medium σ is small, the losses are small and the polarization is not strong. So we can omit terms of lower order of magnitude, these include $\dot{P}_n$, $\sigma\,\dot{\mathcal{E}}_n$, $\ddot{\mathcal{E}}_n$, $\ddot{\phi}_n$, $\sigma\,\dot{\phi}_n$, $\dot{\phi}_n\,\dot{\mathcal{E}}_n$, $\dot{\phi}_n P_n$ etc. This leads to

$$\Omega_n^2\mathcal{E}_n - i\frac{\sigma}{\epsilon_0}\nu_n\mathcal{E}_n \text{-}2i\nu_n\dot{\mathcal{E}}_n - (\nu_n + \dot{\phi}_n)^2\mathcal{E}_n = \frac{\nu_n^2}{\epsilon_0}P_n. \tag{5.4.2}$$

Quality Factor

The condition for oscillation is that the energy lost by the field per unit time is equal to the energy gained per unit time from the medium, i.e.

$Loss = Saturated\ Gain$

Loss can be written in terms of an important property of the cavity called the Quality Factor defined as

$$Q = \Omega\,\frac{energy\ stored\ in\ the\ field}{energy\ lost\ per\ second}.$$

The loss can be defined in terms of the conductivity. The space time dependence of the field $\mathcal{E}$ can be obtained from the rate equation (Eq.5.2.10) without polarization.

$$\frac{\partial^2\mathcal{E}}{\partial z^2} = \mu_0\sigma\frac{\partial\mathcal{E}}{\partial t} + \mu_0\epsilon_0\frac{\partial^2\mathcal{E}}{\partial t^2}. \tag{5.4.3}$$

Assuming spatial dependence as sin $K_n z$ and time dependence as exp $(i\,\Lambda_n t)$ we get

$$-K_n^2 = \mu_0\,\sigma\, i\,\Lambda_n - {\Lambda_n}^2/c^2,$$

$${\Lambda_n}^2 - K_n^2 c^2 - \frac{i\sigma}{\epsilon_0}\Lambda_n = 0.$$

Hence the solution is

$$\Lambda_n = \frac{1}{2}\Big[\frac{i\sigma}{\epsilon_0} \pm \Big(-\big(\frac{\sigma}{\epsilon_0}\big)^2 + 4\,\Omega_n^2\Big)^{\frac{1}{2}}\Big]$$

$$\cong \frac{1}{2}\Big[\frac{i\sigma}{\epsilon_0} \pm 2\,\Omega_n \Big(1 - \frac{1}{\Omega_n}\big(\frac{\sigma}{\epsilon_0}\big)^2\Big)$$

$$\cong \frac{i\sigma}{2\epsilon_0} \pm \Omega_n\,. \tag{5.4.4}$$

Hence the time dependence is $e^{-\frac{\sigma}{2\epsilon_0}t} e^{\pm\, i\Omega_n t}$. The first term describes the attenuation of the field. But in terms of Quality Factor the field in the cavity decays with time as exp $(-\frac{\Omega_n}{Q_n}t)$. Hence

$$\frac{\Omega_n}{2Q_n} = \frac{\sigma}{2\epsilon_0}$$

or,

$$\sigma = \epsilon_0\frac{\Omega_n}{Q_n} \cong \epsilon_0\frac{\nu}{Q_n}. \tag{5.4.5}$$

Since the cavity frequency Ω_n is close to the oscillation frequency $(\nu_n + \dot{\phi}_n)$, we can approximate the cavity frequency as the frequency ν_n, since $\dot{\phi}_n$ is small. We have also neglected the difference in frequencies of individual modes, so that ν is used for ν_n in the expression for conductivity. We also assume

$$\nu_n^2 P_n = \nu\nu_n P_n$$

and

$$\Omega_n^2 - (\nu_n + \dot{\phi}_n)^2 \cong 2\nu_n(\Omega_n - \nu_n - \dot{\phi}_n)\,. \tag{5.4.6}$$

Hence, we have from Eq. (5.4.2)

$$2\nu_n(\Omega_n - \nu_n - \dot{\phi}_n)\mathcal{E}_n - i\frac{\nu\nu_n}{Q_n}\mathcal{E}_n - 2i\nu_n\dot{\mathcal{E}}_n = \frac{1}{\epsilon_0}\nu\nu_n P_n. \tag{5.4.7}$$

Equating the real and imaginary parts we have,`

$$\dot{\mathcal{E}}_n + \frac{\nu}{2Q_n}\mathcal{E}_n = -\frac{1}{2\epsilon_0}\nu\, Im\,(P_n), \tag{5.4.8}$$

$$\Omega_n - (\nu_n + \dot{\phi}_n) = \frac{1}{2\epsilon_0\mathcal{E}_n}\nu\, Re\,(P_n). \tag{5.4.9}$$

These two equations are known as Lamb's Self-consistency equations. These equations may be considered as the basic working equations for the operation of a laser.

Physical Interpretation

It is clear that the polarization occurs because of the laser medium. In absence of an active medium $P_n = 0$. From Eq. (5.4.8) we can write

$$2\mathcal{E}_n\dot{\mathcal{E}}_n = -\frac{\nu}{Q_n}\mathcal{E}_n{}^2$$

Hence

$$\mathcal{E}_n{}^2 = Ae^{-\frac{\nu}{Q_n}t} \tag{5.4.10}$$

Thus, the square of mode amplitude or intensity decreases exponentially in time at a constant rate $\frac{\nu}{Q_n}$. The second equation Eq. (5.4.9) implies that in absence of the medium the mode oscillation frequency is equal to the cavity frequency.

The complex polarization can be written in terms of susceptibility as

$$\boldsymbol{P}_n = \epsilon_0\chi_n\mathcal{E}_n = \epsilon_0\,(\chi'_n + i\chi''_n)\,\mathcal{E}_n \tag{5.4.11}$$

The susceptibility depends both on the mode amplitude and oscillation frequency. Substituting in Eq. (5.4.9) and (5.4.10) we have

$$\dot{\mathcal{E}}_n + \frac{\nu}{2Q_n}\mathcal{E}_n = -\frac{1}{2}\nu\,\chi''_n\mathcal{E}_n\,, \tag{5.4.12}$$

$$\Omega_n - (\nu_n + \dot{\phi}_n) = \frac{1}{2}\nu\,\chi'_n. \tag{5.4.13}$$

The energy per unit volume for a given mode is proportional to the square of the mode amplitude

$$I_n \propto \mathcal{E}_n^2$$

Hence from Eq. (5.4.12)

$$\dot{I}_n = -\frac{\nu}{Q_n}I_n - \nu\,\chi_n''\,I_n. \tag{5.4.14}$$

Hence, energy change per second= − (cavity loss per second) + (gain in the medium per second).

The first term with $\frac{\nu}{Q_n}$ represents the energy loss in the cavity. With $\chi''_n < 0$, the second term $-\nu\,\chi''_n$ represents gain from the medium. Thus the energy change is the difference between the gain from the medium and loss from the cavity. If the medium is sufficiently saturated by the gain parameter χ''_n so as to equal the cavity loss, the steady state condition is achieved. In steady state $\dot{I}_n = 0$, we obtain the condition for laser operation as

$$Saturated\ Gain = Loss.$$

The frequency equation shows that the oscillation frequency $\nu_n + \dot{\phi}_n$ is shifted from the passive cavity frequency Ω_n by a magnitude $-\frac{1}{2}\nu\,\chi'_n$. We can obtain the refractive index of the medium as

$$\eta = \frac{\Omega_n}{(\nu_n + \dot{\phi}_n)} = \frac{\Omega_n}{\Omega_n - \frac{1}{2}\nu\,\chi'_n}. \tag{5.4.15}$$

Since in a dilute medium $\nu \cong \Omega_n$ and χ'_n is small, we get

$$\eta = \frac{1}{1-\frac{1}{2}\chi'_n} \cong 1 + \frac{1}{2}\chi'_n \tag{5.4.16}$$

Such a formula can also be obtained from elementary electromagnetic theory.

5.5 Mode Polarization for two level atoms in a cavity

We consider the laser medium as consisting of two level atoms that are homogeneously broadened (Fig. 5.2). An atom that is excited to the level b at time t_0 and place z may be described by the density matrix $\rho\,(b, z, t, t_0)$ at time t. Atoms are excited to the state b at the rate $\lambda_b(z, t_0)$ per unit time per unit volume. The level populations decay at the rates γ_b and γ_a, respectively and the non-diagonal matrix elements decay at the rate γ. We assume that the lower state a is not the ground state of the atom. We also assume that in the presence of a field, there can be emission from the upper level to the lower level. The macroscopic polarization of the medium at a place z and at time t is described by the contribution of all atoms regardless of their initial states and the time of excitation, so

$$P\,(z,t) = \sum_{n=a,b} \int_{-\infty}^{t} dt_0\, \lambda_n(z,t_0)\, \langle er \rangle$$

$$= \mu \sum_{n=a,b} \int_{-\infty}^{t} dt_0\, \lambda_n(z,t_0)\, \rho_{ab}\,(n, z, t_0, t) \;+ \text{c.c.} \tag{5.5.1}$$

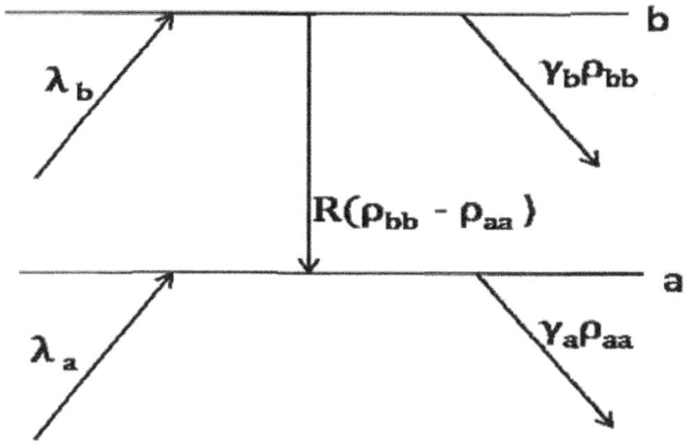

Fig. 5.2 A two level atom showing excitations to and decay from the two levels. It also shows a stimulated emission from the upper level to the lower level.

The quantity $\langle er \rangle = \mu\, \rho_{ab}\,(n, z, t_0, t)$ represents the ensemble average over many atoms. We have written μ for the non-diagonal element μ_{ba}, this is the only non-zero element for the dipole moment matrix of the two-level atom. We can also define a population matrix for the density matrix as

$$\rho\,(z,t) = \sum_{n=a,b} \int_{-\infty}^{t} dt_0\, \lambda_n(z,t_0)\, \rho\,(n, z, t_0, t). \tag{5.5.2}$$

This includes summation over initial states and all times of excitation. Hence the macroscopic polarization is

$$P(z,t) = \mu \rho_{ba}(z,t) + \text{c.c.} \tag{5.5.3}$$

The equations of motion can be obtained by differentiating $\rho\ (z,t)$ with respect to time.

$$\frac{d\rho\ (z,t)}{dt} = \sum_{n=a,b} \lambda_n(z,t)\, \rho\ (n,z,t,t) + \sum_{n=a,b} \int_{-\infty}^{t} dt_0\, \lambda_n(z,t_0)\, \dot{\rho}\ (n,z,t_0,t), \tag{5.5.4}$$

since the first term is

$$\rho_{ij}\ (n,z,t,t) = \delta_{in}\delta_{jn}\,. \tag{5.5.5}$$

The first term is equivalent to the diagonal matrix with elements λ_a and λ_b. Hence the equations of motion of the matrix elements can be written using Eq. (3.3.7), (3.3.8) and (3.3.9) as

$$\dot{\rho}_{aa} = \lambda_a - \Upsilon_a \rho_{aa} + \left(-\frac{i}{\hbar} V_{ab}\rho_{ba} + c.c.\right), \tag{5.5.6}$$

$$\dot{\rho}_{bb} = \lambda_b - \Upsilon_b \rho_{bb} + \left(-\frac{i}{\hbar} V_{ba}\rho_{ab} + c.c.\right) \tag{5.5.7}$$

and

$$\dot{\rho}_{ba} = -(i\omega+\gamma)\, \rho_{ba} + \frac{i}{\hbar} V_{ba}(z,t)(\rho_{bb} - \rho_{aa}). \tag{5.5.8}$$

We can obtain the Fourier components of the polarization from the expressions of Eq. (5.3.4) and (5.5.3)

$$\sum_m P_m\ (t) \exp\{-i\,(\nu_m t + \phi_m)\} \exp\{i\,(\nu_n t + \phi_n)\} \int_0^L dz U_n^*\ (z) U_m\ (z) =$$
$$2\exp\{i\,(\nu_n t + \phi_n)\}\, \mu \int_0^L dz U_n^*\ (z)\, \rho_{ba}\ (z,t) \tag{5.5.9}$$

or,

$$P_m\ (t)\delta_{mn} \int_0^L dz\, |U_n(z)|^2 = 2\mu \exp\{i\,(\nu_n t + \phi_n)\} \int_0^L dz U_n^*\ (z)\ \rho_{ba}(z,t)$$

or,

$$P_n\ (t) = \frac{2\mu}{N_c} \exp\{i\,(\nu_n t + \phi_n)\} \int_0^L dz U_n^*\ (z)\, \rho_{ba}\ (z,t). \tag{5.5.10}$$

where we have written the normalization constant as

$$N_c = \int_0^L dz\, |U_n(z)|^2\,. \tag{5.5.11}$$

5.6 Rate equation approximation

In order to use the self-consistency equation we have to use the normal mode polarization. For this purpose we have to determine the density matrix elements. We shall do it by using single mode rate equation approximation.

Multiplying both sides of Eq. (5.5.8) by $e^{(i\omega+\gamma)t'}$ we get

$$\dot{\rho}_{ba} e^{(i\omega+\gamma)t'} + (i\omega+\gamma)\, \rho_{ba} e^{(i\omega+\gamma)t'} = \frac{i}{\hbar} V_{ba}(z,t')(\rho_{bb}\ (z,t') - \rho_{aa}(z,t'))\ e^{(i\omega+\gamma)t'}$$

or, we can write

$$\frac{d}{dt'}\left[\rho_{ba}(z,t)e^{(i\omega+\gamma)t'}\right] = \frac{i}{\hbar}V_{ba}(z,t')[(\rho_{bb}(z,t') - \rho_{aa}(z,t')]\,e^{(i\omega+\gamma)t'}.$$

Integrating from -∞ to t we get

$$\rho_{ba}(z,t)e^{(i\omega+\gamma)t} = \frac{i}{\hbar}\int_{-\infty}^{t} dt'\, V_{ba}(z,t')[(\rho_{bb}(z,t') - \rho_{aa}(z,t')]\,e^{(i\omega+\gamma)t'}$$

or,

$$\rho_{ba}(z,t) = \frac{i}{\hbar}\int_{-\infty}^{t} dt'\, V_{ba}(z,t')e^{-(i\omega+\gamma)(t-t')}[(\rho_{bb}(z,t') - \rho_{aa}(z,t')]. \qquad (5.6.1)$$

The perturbation energy $V_{ba}(z,t) = -\mu\,\mathcal{E}(z,t)$ for a single mode in the RWA can be written as (using Eq. 5.3.5)

$$V_{ba}(z,t) = -\frac{1}{2}\,\mu\mathcal{E}_n(t)\exp\{-i(\nu_n t + \phi_n)\}U_n(z). \qquad (5.6.2)$$

After substituting the expression for $V_{ba}(z,t)$ we can carry out the time integration. We assume that the mode amplitude $\mathcal{E}_n$, the phase ϕ_n and the population difference $[(\rho_{bb}(z,t') - \rho_{aa}(z,t')]$ terms do not vary appreciably in a time of the order of $1/\gamma$, hence they can be factored out of the integral. These assumptions lead to the rate equations for atomic populations and we call this the Rate Equation Approximation.

$$\rho_{ba}(z,t) = \frac{-i\mu}{2\hbar}\int_{-\infty}^{t} dt'\; \mathcal{E}_n(t')\exp\{-i(\nu_n t' + \phi_n)\}U_n(z)e^{-(i\omega+\gamma)(t-t')}(\rho_{bb}(z,t') - \rho_{aa}(z,t')]$$

$$= \frac{-i\mu\mathcal{E}_n}{2\hbar}e^{-i\phi_n}U_n(z)[(\rho_{bb}(z,t) - \rho_{aa}(z,t)] \quad e^{-(i\omega+\gamma)t}\int_{-\infty}^{t} dt'\, e^{[i(\omega-\nu_n)+\gamma)t'},$$

$$\rho_{ba}(z,t) = \frac{-i\mu\mathcal{E}_n}{2\hbar}e^{-i(\nu_n t+\phi_n)}U_n(z)\frac{(\rho_{bb}(z,t)-\rho_{aa}(z,t)}{i(\omega-\nu_n)+\gamma}. \qquad (5.6.3)$$

Substituting the rate equation for ρ_{bb} in Eq. 5.5.7 we get

$$\dot{\rho}_{bb} = \lambda_b - \Upsilon_b\rho_{bb} + \Big(\frac{i}{2\hbar}\,\mu\mathcal{E}_n(t)\exp\{-i(\nu_n t + \phi_n)\}U_n(z)\frac{(i\mu\mathcal{E}_n)}{2\hbar}e^{i(\nu_n t+\phi_n)}U_n^*(z)\frac{(\rho_{bb}(z,t) - \rho_{aa}(z,t)}{-i(\omega-\nu_n)+\gamma} + C.C.\Big)$$

$$= \lambda_b - \Upsilon_b\rho_{bb} - \Big(\frac{\mu\mathcal{E}_n}{2\hbar}\Big)^2[U_n(z)]^2[\rho_{bb}(z,t) - \rho_{aa}(z,t)]\frac{2\gamma}{(\omega-\nu_n)^2+\gamma^2}$$

$$\dot{\rho}_{bb} = \lambda_b - \Upsilon_b\rho_{bb} - R(z)\;(\rho_{bb}(z,t) - \rho_{aa}(z,t). \qquad (5.6.4)$$

The rate constant R is defined as

$$R(z) = \frac{1}{2\gamma}\Big(\frac{\mu\mathcal{E}_n}{\hbar}\Big)^2[U_n(z)]^2\mathcal{L}(\omega-\nu_n). \qquad (5.6.5)$$

The dimensionless Lorentzian is defined as

$$\mathcal{L}(\omega-\nu_n) = \frac{\gamma^2}{(\omega-\nu_n)^2+\gamma^2}. \qquad (5.6.6)$$

Similarly we can get

$$\dot{\rho}_{aa} = \lambda_a - \Upsilon_a\rho_{aa} + R(z)\;(\rho_{bb}(z,t) - \rho_{aa}(z,t). \qquad (5.6.7)$$

In the steady state $\dot{\rho}_{aa} = \dot{\rho}_{bb} = 0$, this leads to

$$\lambda_b = \Upsilon_b \rho_{bb} + R(z)\ (\rho_{bb}(z,t) - \rho_{aa}(z,t),$$

$$\lambda_a = \Upsilon_a \rho_{aa} - R(z)\ (\rho_{bb}(z,t) - \rho_{aa}(z,t),$$

$$\frac{\lambda_b}{\Upsilon_b} - \frac{\lambda_a}{\Upsilon_a} = [\rho_{bb}(z,t) - \rho_{aa}(z,t)][\,1 + R(z)(\frac{1}{\Upsilon_b} + \frac{1}{\Upsilon_a})\,],$$

$$[\rho_{bb}(z,t) - \rho_{aa}(z,t)] = \frac{N(z)}{1+R(z)/R_s}, \qquad (5.6.8)$$

where,

$$N(z) = \frac{\lambda_b}{\Upsilon_b} - \frac{\lambda_a}{\Upsilon_a} \qquad (5.6.9)$$

is the unsaturated population difference, when $\mathcal{E}_n = 0$ or, $R(z) = 0$ and

$$R_s = \frac{\Upsilon_b \Upsilon_a}{2\gamma_{ab}}, \qquad (5.6.10)$$

R_s is called the saturation parameter with the decay constant defined as $\gamma_{ab} = (\Upsilon_b + \Upsilon_a)/2$.

Substituting the expression of Eq. (5.6.8) we can obtain the non-diagonal element $\rho_{ba}(z,t)$.

5.7 Hole Burning

Thus the population difference is given by the unsaturated population difference $N(z)$ divided by the factor $1 + R/R_s$ that increases with the increase of the electric field. If $R(z) = R_s$, the population difference is half of the unsaturated value. Because of the normal mode oscillation term $[U_n(z)]^2 = \sin^2 K_n z$ the population difference has also a sinusoidal dependence. At the nodes of the field, $R(z) = 0$, hence the population difference $[\rho_{bb}(z,t) - \rho_{aa}(z,t)]$ assumes the zero-field value. Since the maximum value of $[U_n(z)]^2$ is $+1$, there is always decrease from the zero field value. For a particular value of the electric field intensity the population difference oscillates between a maximum and a minimum value. A plot (Fig. 5.3) of the curve of $[\rho_{bb}(z,t) - \rho_{aa}(z,t)]/\,N(z)$

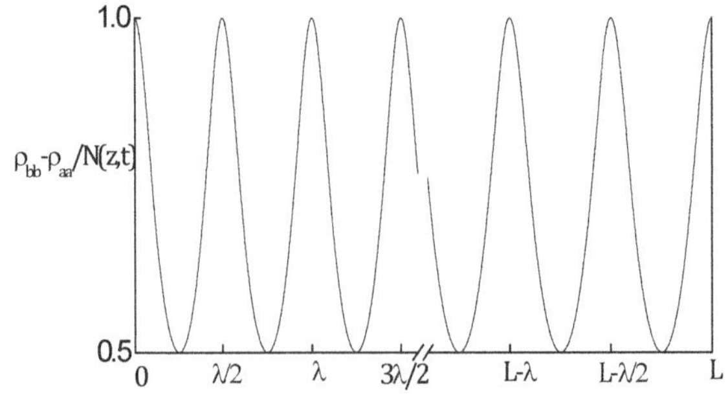

Fig.5.3 Hole burning on the population difference curve caused by the electromagnetic field in a linear cavity of length L. L is a multiple of $\lambda/2$.

shows that the electric field burns holes in the population difference curve along the z-direction. The maxima and minima are one half wavelength apart since $\frac{R(z)}{R_s} \propto \sin^2 K_n z$. This is called Hole Burning. This phenomenon of hole burning is also common in laser spectroscopy as it was discussed in Chapter 4.

5.8 Non-linear Polarization

We can obtain the single mode polarization of Eq. (5.5.10) after using Eq. (5.6.3) and (5.6.8)

$$P_n(t) = \frac{-i\mu^2}{N_c}\frac{\mathcal{E}_n}{\hbar}\int_0^L dz\,|U_n(z)|^2 \frac{\frac{N(z)}{1+R(z)/R_s}}{i(\omega-\nu_n)+\gamma}$$

$$= \frac{-\mu^2}{N_c}\frac{\mathcal{E}_n}{\hbar}\frac{(\omega-\nu_n)+i\gamma}{(\omega-\nu_n)^2+\gamma^2}\int_0^L dz\,|U_n(z)|^2 \frac{N(z)}{1+R(z)/R_s}. \quad (5.8.1)$$

The integral over z depends on the unsaturated population difference $N(z)$ and also the sinusoidal z-dependence of the rate constant $R(z)$. We assume that $N(z)$ does not vary appreciably over the length and also assume that $\mathcal{E}_n$ is small so that the term $\frac{1}{1+R(z)/R_s}$ can be expanded in power series. Hence the first order term in the complex polarization can be written as the first term with $R(z)=0$. Hence the first order polarization is

$$P_n^{(1)}(t) = \frac{-\mu^2\mathcal{E}_n}{\hbar}\widetilde{N}\frac{(\omega-\nu_n)+i\gamma}{(\omega-\nu_n)^2+\gamma^2}, \quad (5.8.2)$$

where,

$$\widetilde{N} = \frac{1}{N_c}\int_0^L dz\,|U_n(z)|^2 N(z). \quad (5.8.3)$$

The term is linear in $\mathcal{E}_n$ and contains no saturation term. It is valid for the threshold condition and is true for the minimum excitation for which oscillation can begin.

In the standing wave case

$$|U_n(z)|^2 = \sin^2 K_n z = \frac{1}{2}(1 - \cos 2K_n z).$$

Hence,

$$N_c = \int_0^L dz\,|U_n(z)|^2 = L/2. \quad (5.8.4)$$

$$\widetilde{N} = \frac{1}{N_c}\int_0^L dz\,|U_n(z)|^2 N(z) = \frac{1}{L}\int_0^L dz\,N(z). \quad (5.8.5)$$

We have assumed that $N(z)$ is slowly varying over a wavelength. The linear approximation assumed above is valid only for the threshold condition.

By expansion of the denominator and retaining only the next higher order term or the term that is linear in $R(z)$ we can get the next higher order term in polarization. Since $R(z)$ contains a term $|U_n(z)|^2$, the integral in Eq. (5.8.1) will involve a term $|U_n(z)|^4$.

$$|U_n(z)|^4 = ¼(1 - \cos 2K_n z)^2$$

$$= \frac{3}{8} - ½\cos 2K_n z + \frac{1}{8}\cos 4K_n z.$$

Hence,

$$\frac{1}{N_c}\int_0^L dz\,|U_n(z)|^4 N(z) = \frac{1}{L/2}\int_0^L dz\,\frac{3}{8}N(z) = \frac{3}{4}\widetilde{N}. \quad (5.8.6)$$

Hence, the third order term in polarization (from Eq. (5.6.5) and (5.8.1)) is

$$P_n^{(3)}(t) = \frac{3}{4}\frac{\mu^2\mathcal{E}_n}{\hbar}\tilde{N}\frac{(\omega-\nu_n)+i\gamma}{(\omega-\nu_n)^2+\gamma^2}\frac{1}{2\gamma R_s}\left(\frac{\mu\mathcal{E}_n}{\hbar}\right)^2\mathcal{L}(\omega-\nu_n)\,.$$

Total polarization up to third order term is

$$P_n(t) = P_n^{(1)}(t) + P_n^{(3)}(t)$$

$$= -\frac{\mu^2\mathcal{E}_n}{\hbar}\tilde{N}\frac{(\omega-\nu_n)+i\gamma}{(\omega-\nu_n)^2+\gamma^2}\left[1-\frac{3}{2}\frac{\gamma_{ab}\gamma I_n}{(\omega-\nu_n)^2+\gamma^2}\right]. \tag{5.8.7}$$

where, the dimensionless intensity is defined as

$$I_n = \frac{1}{2\Upsilon_b\Upsilon_a}\left(\frac{\mu\mathcal{E}_n}{\hbar}\right)^2. \tag{5.8.8}$$

Since the second term in Eq. (5.8.7) is small, we can approximate as

$$P_n(t) = -\frac{\mu^2\mathcal{E}_n}{\hbar}\tilde{N}\frac{(\omega-\nu_n)+i\gamma}{[(\omega-\nu_n)^2+\gamma^2][1+\frac{3}{2}\frac{\gamma_{ab}\gamma I_n}{(\omega-\nu_n)^2+\gamma^2}]}$$

$$= -\frac{\mu^2\mathcal{E}_n}{\hbar}\tilde{N}\frac{(\omega-\nu_n)+i\gamma}{(\omega-\nu_n)^2+\gamma^2[1+\frac{3\gamma_{ab}I_n}{2\gamma}]}. \tag{5.8.9}$$

At resonance, the polarization is decreased by a factor $1+\frac{3}{2}\frac{\gamma_{ab}I_n}{\gamma}$. This additional term thus leads to power broadening or saturation broadening. The effective line width can be given approximately as

$$\gamma_{eff} = \gamma\left[1+\frac{3}{2}\frac{\gamma_{ab}I_n}{\gamma}\right]^{1/2} \tag{5.8.10}$$

From equation (5.4.11) and Eq. (5.8.9) we can write the real and imaginary parts of the susceptibility as

$$\chi_n' = -\frac{\mu^2}{\hbar\epsilon_0}\tilde{N}\frac{(\omega-\nu_n)}{(\omega-\nu_n)^2+\gamma^2[1+\frac{3\gamma_{ab}I_n}{2\gamma}]}, \tag{5.8.11}$$

$$\chi_n'' = -\frac{\mu^2}{\hbar\epsilon_0}\tilde{N}\frac{\gamma}{(\omega-\nu_n)^2+\gamma^2[1+\frac{3\gamma_{ab}I_n}{2\gamma}]}. \tag{5.8.12}$$

A plot of both χ_n' and χ_n'' in Fig. 5.4 shows the variation with $\omega-\nu_n$. It is shown that χ_n'' has a negative peak at $\nu_n=\omega$. χ_n' shows the dispersive behavior.

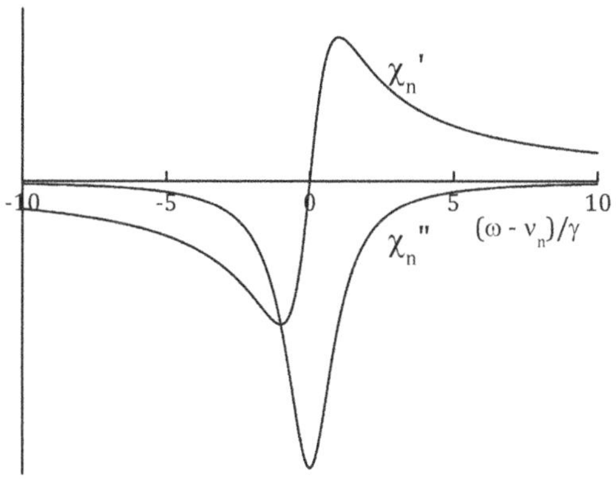

Fig. 5.4 Plot of real and imaginary parts of susceptibility vs. detuning.

5.9 Intensity and Frequency of a Single Mode Laser

From the Self-Consistency Equation (Eq. (5.4.8)) and Eq.(5.8.7) for non-linear polarization, we get

$$\dot{\mathcal{E}}_n = -\frac{\nu}{2Q_n}\mathcal{E}_n + \frac{1}{2\epsilon_0}\nu\frac{\mu^2\mathcal{E}_n}{\hbar}\widetilde{N}\frac{\gamma}{(\omega-\nu_n)^2+\gamma^2}\left[1-\frac{3}{2}\frac{\gamma_{ab}\gamma I_n}{(\omega-\nu_n)^2+\gamma^2}\right]$$

$$=\mathcal{E}_n\left\{-\frac{\nu}{2Q_n}+F_1\mathcal{L}(\omega-\nu_n)-F_1\frac{3}{2}\frac{\gamma_{ab}I_n}{\gamma}[\mathcal{L}(\omega-\nu_n)]^2\right\}, \quad (5.9.1)$$

where,

$$F_1 = \frac{\nu}{2\epsilon_0\gamma}\frac{\mu^2}{\hbar}\widetilde{N} \quad (5.9.2)$$

Hence, we can write

$$\dot{\mathcal{E}}_n = \mathcal{E}_n(\alpha_n - \beta_n I_n), \quad (5.9.3)$$

where,

$$\alpha_n = F_1\mathcal{L}(\omega-\nu_n) - \frac{\nu}{2Q_n}, \quad (5.9.4)$$

$$\beta_n = F_1\frac{3}{2}\frac{\gamma_{ab}}{\gamma}[\mathcal{L}(\omega-\nu_n)]^2. \quad (5.9.5)$$

α_n is the net gain coefficient and β_n is the self-saturation coefficient. Since I_n is the dimensionless intensity defined by Eq. (5.8.8), we multiply Eq. (5.9.3) by $\frac{\mathcal{E}_n}{\Upsilon_b\Upsilon_a}(\frac{\mu}{\hbar})^2$ to obtain the rate equation for I_n as

$$\dot{I_n} = 2I_n(\alpha_n - \beta_n I_n). \quad (5.9.6)$$

The intensity increases by the net gain obtained from the difference of gain from atomic transitions and cavity loss as expressed by the two terms of α_n in Eq. (5.9.4). The increase in gain is compensated by the saturation term given by Eq. (5.9.5).

Mode pulling and mode pushing

In order to determine the oscillation frequency we have from Eq. (5.4.9)

$$(\nu_n + \dot{\phi}_n) = \Omega_n - \frac{1}{2\epsilon_0 \mathcal{E}_n} \nu \, Re\,(P_n)$$

$$= \Omega_n + \frac{\nu \mu^2}{2\hbar \epsilon_0} \tilde{N} \frac{(\omega - \nu_n)}{(\omega - \nu_n)^2 + \gamma^2} \left[1 - \frac{3}{2} \frac{\gamma_{ab} \gamma I_n}{(\omega - \nu_n)^2 + \gamma^2}\right]$$

or,

$$(\nu_n + \dot{\phi}_n) = \Omega_n + \sigma_n - \rho_n I_n\,, \tag{5.9.7}$$

where,

$$\sigma_n = F_1 \frac{\omega - \nu_n}{\gamma} \mathcal{L}\,(\omega - \nu_n) \tag{5.9.8}$$

is the mode pulling coefficient and

$$\rho_n = F_1 \frac{\omega - \nu_n}{\gamma} \frac{3}{2} \frac{\gamma_{ab}}{\gamma} [\mathcal{L}\,(\omega - \nu_n)]^2 \tag{5.9.9}$$

is the mode-pushing coefficient. In the frequency equation the oscillation frequency deviates from the cavity frequency by two terms. The first term arising from the first order polarization term pulls the frequency away from the cavity frequency. This frequency pulling is over-estimated by the unsaturated term, σ_n; this is corrected by the third order polarization term, $\rho_n I_n$ that pushes it further towards the cavity frequency. Normally, the first term dominates over the second term and there is a frequency pulling. Dropping the phase term $\dot{\phi}_n$ we can write

$$\nu_n = \Omega_n + \sigma_n - \rho_n I_n$$

$$= \Omega_n + F_1 \frac{\omega - \nu_n}{\gamma} \mathcal{L}\,(\omega - \nu_n)\left[1 - \frac{3}{2} \frac{\gamma_{ab}}{\gamma} I_n \mathcal{L}\,(\omega - \nu_n)\right].$$

It can be easily shown that

$$\nu_n = \frac{\Omega_n + S\omega}{1 + S}\,, \tag{5.9.10}$$

where,

$$S = F_1 \frac{1}{\gamma} \mathcal{L}\,(\omega - \nu_n)\left[1 - \frac{3}{2} \frac{\gamma_{ab}}{\gamma} I_n \mathcal{L}\,(\omega - \nu_n)\right] \tag{5.9.11}$$

is the stabilizing factor.

In the steady state, the intensity equation leads to
$\dot{I}_n = 0$, hence

$$\alpha_n = \beta_n I_n.$$

$$F_1\mathcal{L}(\omega - \nu_n) - \frac{\nu}{2Q_n} = F_1\frac{3}{2}\frac{\gamma_{ab}}{\gamma}[\mathcal{L}(\omega - \nu_n)]^2 I_n$$

or,

$$\frac{\nu}{2Q_n} = F_1\mathcal{L}(\omega - \nu_n)[1 - \frac{3}{2}\frac{\gamma_{ab}}{\gamma}\mathcal{L}(\omega - \nu_n)I_n] = \gamma S$$

or,

$$S = \frac{\nu}{2Q_n\gamma}. \quad (5.9.12)$$

Hence,

$$\nu_n = \frac{\gamma\Omega_n + \frac{\nu}{2Q_n}\omega}{\gamma + \frac{\nu}{2Q_n}}. \quad (5.9.13)$$

This equation shows the frequency as the average of the cavity frequency and the atomic resonance frequency with different weight factors γ and $\frac{\nu}{2Q_n}$. For a high quality cavity, we have $\frac{\nu}{2Q_n} \ll \gamma$. Hence, $\nu_n \cong \Omega_n$, so the frequency is close to the cavity frequency, although pulled somewhat towards the resonance frequency. When γ is small and the cavity has low quality the oscillation frequency is close to the resonance frequency ω.

The index of refraction can be obtained as

$$\eta(\nu_n) = \frac{\Omega_n}{\nu_n} = (1 + \frac{\nu}{2Q_n\gamma})(1 + \frac{\nu}{2Q_n\gamma}\frac{\omega}{\Omega_n})^{-1}$$

$$= 1 + \frac{\nu}{2Q_n\gamma}(1 - \frac{\omega}{\Omega_n}). \quad (5.9.14)$$

In the above expression, higher order terms are omitted. A plot of $\eta(\nu_n - \omega) - 1$ against $(\nu_n - \omega)$ in MHz (Fig. 5.5) shows the dispersive behavior. We have assumed $\frac{\nu}{2Q_n} \cong 0.01\gamma$ and $\nu_n \cong \Omega_n$. Hence $\eta(\nu_n - \omega) - 1 = 0.01(\nu_n - \omega)/\nu_n$. Thus, the refractive index is unity at resonance and is greater than one or less than one in off-resonance cases.

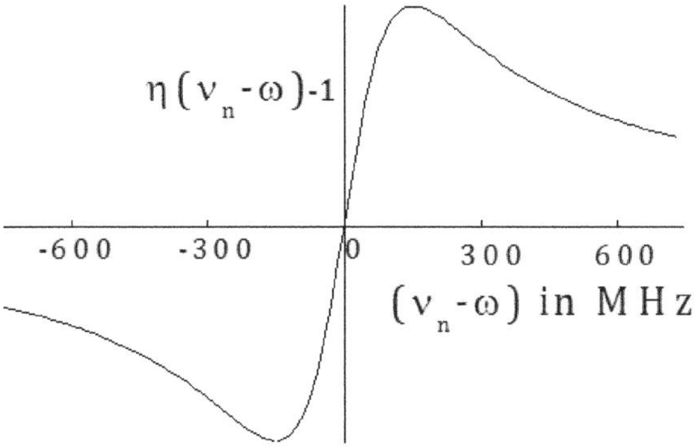

Fig. 5.5 The index of refraction plotted against detuning for a gain medium with $\tilde{N} > 0$. The curve is evaluated near threshold with $\frac{\nu}{2Q_n\gamma} = 0.01$

Threshold condition

We can obtain the threshold condition by setting the net gain coefficient at resonance to zero. This leads to

$$\frac{\nu}{2\epsilon_0\gamma}\frac{\mu^2}{\hbar}\widetilde{N}_T = \frac{\nu}{2Q_n},$$

$\widetilde{N}_T$ is the threshold value of $\widetilde{N}$ at resonance.

$$\widetilde{N}_T = \frac{\hbar}{Q_n}\frac{\epsilon_0\gamma}{\mu^2}. \tag{5.9.15}$$

We define the relative excitation as

$$\mathbb{N} = \frac{\widetilde{N}}{\widetilde{N}_T} \tag{5.9.16}$$

In this case, the first order factor becomes

$$F_1 = \frac{\nu}{2Q_n}\mathbb{N}$$

At steady state,

$$I_n = \frac{\alpha_n}{\beta_n} = \frac{\mathcal{L}(\omega-\nu_n)-\frac{\nu}{2F_1Q_n}}{\frac{3\gamma_{ab}}{2\gamma}[\mathcal{L}(\omega-\nu_n)]^2} = \frac{\mathcal{L}(\omega-\nu_n)-1/\mathbb{N}}{\frac{3\gamma_{ab}}{2\gamma}[\mathcal{L}(\omega-\nu_n)]^2} \tag{5.9.17}$$

In order to calculate the dimensionless intensity, we have to evaluate the Lorentzian for a certain detuning. For this purpose, we must know the frequency ν_n, given by Eq. (5.9.10) and (5.9.11). But the frequency itself is a function of the intensity. Hence the equation cannot be solved exactly. We can follow an iterative method. In the first order approximation, we can use the cavity frequency as the oscillation frequency and calculate the intensity. The calculated first order intensity can be used to calculate the corrected first order frequency ν_n. This value may be used in a second order iteration. This numerical procedure can be repeated a few times and the result converges after a few iterations. The dimensionless intensity is plotted against detuning in Fig. 5.6 for homogeneously broadened cases. The relative excitations $\mathbb{N}$ are increased from the lower curve to the upper curves in the range 1.0 to 1.2. The intensity increases with relative excitations.

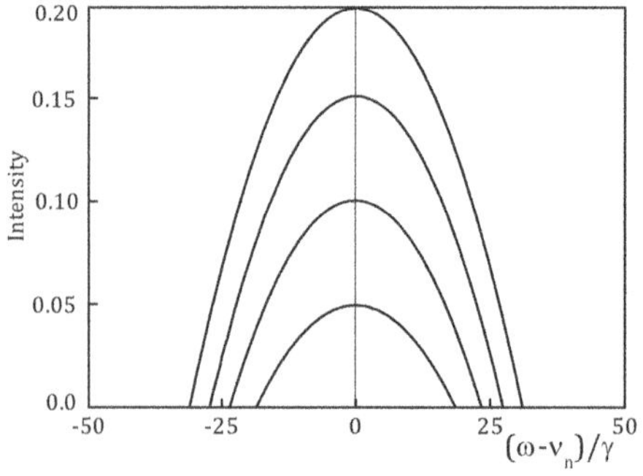

Fig. 5.6 Steady state dimensionless intensity plotted against detuning.

5.10 Two mode operation

Two-mode operation of a laser is the simplest case of multimode operation. We can present an approximate treatment by assuming a weak coupling between the modes. The coupling leads to a cross-saturation in addition to the self-saturation of the individual modes. Direct saturation source provides competition for the gain between the sources. Population pulsation of the nonlinear gain medium provides pulsations at inter-mode frequency.

When two modes are tuned above threshold we can write, following Eq. (5.9.6)),

$$\dot{I}_1 = 2I_1(\alpha_1 - \beta_1 I_1 - \varsigma_{12} I_2), \tag{5.10.1}$$

$$\dot{I}_2 = 2I_2(\alpha_2 - \beta_2 I_2 - \varsigma_{21} I_1). \tag{5.10.2}$$

$\mathcal{E}_n$, α_n, β_n, I_n (n = 1, 2) are defined earlier. ς_{12} and ς_{21} are the cross-saturation terms that can be derived from the theory of multimode operation to be discussed later. Such equations may be written intuitively as an extension of single mode equation.

For the frequency determining equations we can correspondingly write from Eq. (5.9.7)

$$(\nu_1 + \dot{\phi}_1) = \Omega_1 + \sigma_1 - \rho_1 I_1 - \xi_{12} I_2, \tag{5.10.3}$$

$$(\nu_2 + \dot{\phi}_2) = \Omega_2 + \sigma_2 - \rho_2 I_2 - \xi_{21} I_1. \tag{5.10.4}$$

The additional terms, ξ_{12} and ξ_{21} represent the cross-correlation coefficients.

Stationary states are solutions of the steady state equations, $\dot{I}_1 = \dot{I}_2 = 0$. There can be four possible solutions,

$$I_1 = I_2 = 0,$$
$$I_1 = 0,\ I_2 \neq 0,$$
$$I_1 \neq 0,\ I_2 = 0,$$
$$I_1 \neq 0,\ I_2 \neq 0.$$

We must consider solutions which are both physical and non-negative. If both the net gain coefficients α_i $(i = 1,2)$ are negative, there will be no oscillation. If the gain of one mode is negative and that of the other is positive, the mode with positive gain will oscillate with steady state value like α_n/β_n, provided that mode coupling coefficients are positive. If both the gain coefficients are positive, a detailed analysis is required.

We can find the stationary state intensities, I_1^0 and I_2^0. If neither of the intensities is zero, the stationary state solution conditions from Eq. (5.10.1) and (5.10.2) require

$$\beta_1 I_1^0 + \varsigma_{12} I_2^0 = \alpha_1, \tag{5.10.5}$$
$$\beta_2 I_2^0 + \varsigma_{21} I_1^0 = \alpha_2. \tag{5.10.6}$$

These equations are linear in intensities and hence can be solved to obtain

$$I_1^0 = \frac{\frac{\alpha_1}{\beta_1} - \alpha_2 \varsigma_{12}/\beta_1\beta_2}{1 - \varsigma_{12}\varsigma_{21}/\beta_1\beta_2}, \tag{5.10.7}$$

$$I_2^0 = \frac{\frac{\alpha_2}{\beta_2} - \alpha_1 \varsigma_{21}/\beta_1\beta_2}{1 - \varsigma_{12}\varsigma_{21}/\beta_1\beta_2}. \tag{5.10.8}$$

By defining

$$\alpha_1{}^{eff} = \alpha_1 - \frac{\alpha_2 \varsigma_{12}}{\beta_2} \tag{5.10.9}$$

and

$$\alpha_2{}^{eff} = \alpha_2 - \alpha_1 \varsigma_{21}/\beta_1 \tag{5.10.10}$$

as the effective gains and

$$M = \varsigma_{12}\varsigma_{21}/\beta_1\beta_2 \tag{5.10.11}$$

as the mode coupling constant we have

$$I_1^0 = \frac{\frac{\alpha_1{}^{eff}}{\beta_1}}{1-M}, \tag{5.10.12}$$

$$I_2^0 = \frac{\frac{\alpha_2{}^{eff}}{\beta_2}}{1-M}. \tag{5.10.13}$$

If the mode coupling coefficient M is small, $\alpha_1{}^{eff} \rightarrow \alpha_1$ and $\alpha_2{}^{eff} \rightarrow \alpha_2$, in this case, the modes oscillate independently with intensities $\frac{\alpha_1}{\beta_1}$ and $\frac{\alpha_2}{\beta_2}$ respectively. If an effective gain coefficient is negative, the solution is unphysical and it should be discarded. When M is smaller than 1, the effective net gain coefficient with positive value will oscillate. Both the nodes having positive effective gain start oscillating. One will try to inhibit the other. If the coupling parameter M is larger than one, we need the effective gain coefficient to be negative so as to have positive intensity. This is not a stable solution.

Two-mode operation is a simplified approach, but it can be effectively used for a ring laser with three mirrors.

5.11 Multimode operation

In section 5.6, we used the Rate Equation Approximation so that the population difference of a single mode varies little in the time interval $1/\gamma$. In case more than one mode is present in the system, the population difference $\rho_{bb}(z,t) - \rho_{aa}(z,t)$ has pulsations at beat frequencies and multiples of inter-mode frequencies. This process is analogous to the Raman Effect. The rate equation approximation is not valid in such cases and we have to use the multimode field as given in Eq. (5.3.5)

$$\mathcal{E}(z,t) = \frac{1}{2}\sum_n \mathcal{E}_n(t) \exp\{-i(\nu_n t + \phi_n)\} U_n(z) + c.c. \tag{5.11.1}$$

The inter-mode coupling terms result from non-linearity in the polarization. We shall assume that the field is not too strong and the electric field amplitude and phase vary little in one atomic lifetime. We can derive the non-linear polarization keeping terms up to third order in magnitude. The complex polarization components can be used in the self-consistency equations to yield the intensity and oscillation frequency. The phase dependent terms play a significant role.

The multimode perturbation energy can be written as

$$V_{ba}(z,t) = -\frac{1}{2}\sum_n \mu\mathcal{E}_n(t) \exp\{-i(\nu_n t + \phi_n)\} U_n(z). \tag{5.11.2}$$

In the zero-th order ($V_{ba} = 0$) the diagonal elements can be obtained by using an integrating factor in Eqs (5.5.6) and (5.5.7)

$$\frac{d\rho_{bb}^{(0)}e^{\gamma_b t}}{dt} = \lambda_b e^{\gamma_b t}.$$

Integrating from $-\infty$ to t we get

$$\rho_{bb}^{(0)}(z,t) = \frac{\lambda_b}{\gamma_b}. \tag{5.11.3}$$

Similarly, for the lower level

$$\rho_{aa}^{(0)}(z,t) = \frac{\lambda_a}{\gamma_a}. \tag{5.11.4}$$

Hence, the zero-th order population difference is

$$N^0(z,t) = \frac{\lambda_b}{\gamma_b} - \frac{\lambda_a}{\gamma_a}. \tag{5.11.5}$$

Using this population difference, we can find the first order non-diagonal density matrix element from Eq. (5.5.8) as

$$\rho_{ba}^{(1)}(z,t) = \frac{i}{\hbar}\int_{-\infty}^{t} dt' e^{-(i\omega+\gamma)(t-t')} V_{ba}(z,t')N^0(z,t')$$

$$= \frac{i\mu}{2\hbar} N^0(z,t) \sum_n \mathcal{E}_n(t) e^{-i(\nu_n t+\phi_n)} U_n(z)$$

$$\times \int_{-\infty}^{t} dt' e^{-[i(\omega-\nu_n)+\gamma](t-t')}$$

$$= \frac{i\mu}{2\hbar} N^0(z,t) \sum_n \mathcal{E}_n(t) e^{-i(\nu_n t+\phi_n)} U_n(z)$$

$$\times \frac{1}{i(\omega-\nu_n)+\gamma}. \tag{5.11.6}$$

In the process of integration, we have assumed that the variables $\mathcal{E}_n$, ϕ_n and N^0 vary little in time $1/\gamma$ and hence are taken out of the integral. The variation is much slow compared to the variation of the oscillatory terms. Substituting this in the rate equation for the upper level, we obtain the diagonal element in the second order

$$\dot{\rho}_{bb}^{(2)} = \lambda_b - \Upsilon_b \rho_{bb} - [\left(\frac{\mu}{2\hbar}\right)^2 N^0(z,t) \sum_\alpha \sum_\beta \mathcal{E}_\alpha \mathcal{E}_\beta$$

$$\times e^{i[(\nu_\alpha-\nu_\beta)t+\phi_\alpha-\phi_\beta]} U_\alpha^* U_\beta \frac{1}{i(\omega-\nu_\beta)+\gamma} + c.c.]. \tag{5.11.7}$$

This equation shows the oscillation at the inter-mode frequency $(\nu_\alpha - \nu_\beta)$, which is not present in the single mode operation. In the case of multimode operation, the summation over α and β ensures the coupling of all the modes and all possible beat frequencies. In order to obtain the second order term we use the integrating factor and the equation looks like

$$\frac{d\rho_{bb}^{(2)}e^{\gamma_b t}}{dt} = \lambda_b e^{\gamma_b t} - [\left(\frac{\mu}{2\hbar}\right)^2 N^0(z,t) \sum_\alpha \sum_\beta \mathcal{E}_\alpha \mathcal{E}_\beta U_\alpha^* U_\beta$$

$$\times \frac{1}{i(\omega-\nu_\beta)+\gamma} e^{i[(\nu_\alpha-\nu_\beta)t+\phi_\alpha-\phi_\beta]} \int_{-\infty}^{t} e^{[-[i(\nu_\alpha-\nu_\beta)+\gamma_b](t-t')]}$$

or,

$$\rho_{bb}^{(2)} = \frac{\lambda_b}{\gamma_b} - [\left(\frac{\mu}{2\hbar}\right)^2 N^0(z,t) \sum_\alpha \sum_\beta \mathcal{E}_\alpha \mathcal{E}_\beta U_\alpha^* U_\beta$$

$$\times e^{i[(\nu_\alpha-\nu_\beta)t+\phi_\alpha-\phi_\beta]} \frac{1}{i(\omega-\nu_\beta)+\gamma} \cdot \frac{1}{i(\nu_\alpha-\nu_\beta)+\gamma_b} + c.c. \quad (5.11.8)$$

In the above we have assumed that the field mode amplitude and phases vary slowly in time for one decay cycle. This is true for a gas laser with large value of γ_b, but not for a laser like Ruby, which has a $\gamma_b \cong 2\pi \times 1\,kHz$. Similarly, we can write from Eq. (5.5.7)

$$\rho_{aa}^{(2)} = \frac{\lambda_a}{\gamma_a} + [\left(\frac{\mu}{2\hbar}\right)^2 N^0(z,t) \sum_\alpha \sum_\beta \mathcal{E}_\alpha \mathcal{E}_\beta U_\alpha^* U_\beta$$

$$\times e^{i[(\nu_\alpha-\nu_\beta)t+\phi_\alpha-\phi_\beta]} \frac{1}{i(\omega-\nu_\beta)+\gamma} \cdot \frac{1}{i(\nu_\alpha-\nu_\beta)+\gamma_a} + c.c. \quad (5.11.9)$$

Eq, (5.11.8) can be written in a complete form by writing the complex conjugate, which results in an expression with α and β interchanged except the term $\frac{1}{i(\omega-\nu_\beta)+\gamma}$. Hence, it can be written as

$$\rho_{bb}^{(2)} = \frac{\lambda_b}{\gamma_b} - [\left(\frac{\mu}{2\hbar}\right)^2 N^0(z,t) \sum_\alpha \sum_\beta \mathcal{E}_\alpha \mathcal{E}_\beta U_\alpha^* U_\beta$$

$$\times e^{i[(\nu_\alpha-\nu_\beta)t+\phi_\alpha-\phi_\beta]} . \frac{1}{i(\nu_\alpha-\nu_\beta)+\gamma_b} [\frac{1}{i(\omega-\nu_\beta)+\gamma}$$

$$+ \frac{1}{i(\nu_\beta-\omega)+\gamma}]. \quad (5.11.10)$$

Hence,

$$\rho_{bb}^{(2)} - \rho_{aa}^{(2)} = N^0(z,t) - \ [\left(\frac{\mu}{2\hbar}\right)^2 N^0(z,t) \sum_\alpha \sum_\beta \mathcal{E}_\alpha \mathcal{E}_\beta$$

$$\times U_\alpha^* U_\beta \, e^{i[(\nu_\alpha-\nu_\beta)t+\phi_\alpha-\phi_\beta]} [\frac{1}{i(\omega-\nu_\beta)+\gamma}$$

$$+ \frac{1}{i(\nu_\beta-\omega)+\gamma}][\frac{1}{i(\nu_\alpha-\nu_\beta)+\gamma_b} + \frac{1}{i(\nu_\alpha-\nu_\beta)+\gamma_a}]. \quad (5.11.11)$$

Substituting the second order correction term in the integral for $\rho_{ba}(z,t)$, we can get the third order correction to the non-diagonal density matrix element as

$$\rho_{ba}{}^{(3)} = \frac{i}{2} [\left(\frac{\mu}{2\hbar}\right)^2 N^0(z,t) \sum_\alpha \sum_\beta \sum_\theta \mathcal{E}_\theta \mathcal{E}_\alpha \mathcal{E}_\beta U_\theta U_\alpha^* U_\beta$$

$$\times \, e^{i[(\nu_\alpha-\nu_\beta-\nu_\theta)t+\phi_\alpha-\phi_\beta-\phi_\theta]} \frac{1}{i(\omega-\nu_\theta+\nu_\alpha-\nu_\beta)+\gamma}$$

$$\times [\frac{1}{i(\omega-\nu_\beta)+\gamma} + \frac{1}{i(\nu_\beta-\omega)+\gamma}][\frac{1}{i(\nu_\alpha-\nu_\beta)+\gamma_b} +$$

$$\frac{1}{i(\nu_\alpha-\nu_\beta)+\gamma_a}]. \quad (5.11.12)$$

For evaluation of polarization (Eq. 5.5.10) in the third order we have to calculate an integral of the type

$$\int_0^L dz U_n^*(z) U_\theta(z) U_\alpha^*(z) U_\beta(z)$$

Since $U_n(z) = \sin K_n z$ we have terms like

$$\sin K_n z \sin K_\theta z \sin K_\alpha z \sin K_\beta z = \frac{1}{8}[\cos\{(K_n - K_\theta + K_\alpha - K_\beta)z\} + \cos\{(K_n - K_\theta - K_\alpha + K_\beta)z\}$$

$$+\cos\{(K_n + K_\theta - K_\alpha - K_\beta)z\}. \quad (5.11.13)$$

In the trigonometric identity, we have retained only terms slowly varying in z, so that only terms with two positive and two negative K values are kept and all terms with more than two positive or negative K values are neglected. These terms vanish in the integral over z. The expression of polarization will also include oscillatory terms like $e^{i((\nu_n+\nu_\alpha-\nu_\beta-\nu_\theta)t}$. Since the mode cannot have high frequency variation, we retain only terms with

$$\nu_n + \nu_\alpha - \nu_\beta - \nu_\theta = 0.$$

Hence, we should have

$$n = \theta + \beta - \alpha. \quad (5.11.14)$$

This assumption will also simplify the product of sine terms. Substituting Eq. (5.11.12) in Eq. (5.5.10) we can obtain the third order polarization term. It may be noted that the complete expression will involve the phase term $e^{i\varphi_{n\theta\alpha\beta}}$ where,

$$\varphi_{n\theta\alpha\beta} = (\nu_n + \nu_\alpha - \nu_\beta - \nu_\theta)t + \phi_n - \phi_\theta + \phi_\alpha - \phi_\beta. \quad (5.11.15)$$

So the polarization including the first order and the third order terms from Eq. (5.5.10), Eq. (5.11. 6) and Eq. (5.11.12) becomes

$$P_n(t) = \sum_\alpha \sum_\beta \sum_\theta \mathcal{E}_\theta \mathcal{E}_\alpha \mathcal{E}_\beta \wp_{n\theta\alpha\beta} e^{i\varphi_{n\theta\alpha\beta}}. \quad (5.11.16)$$

The amplitude function $\wp_{n\theta\alpha\beta}$ results from the integral of Eq. (5.5.10) involving the time independent terms. $N^0(z,t)$ is assumed as a constant.

Using the polarization of Eq. (5.11.16) in Lamb's Self-consistency equation we can determine the multimode amplitude and frequency.

$$\dot{\mathcal{E}}_n + \frac{\nu}{2Q_n}\mathcal{E}_n = -\frac{\nu}{2\epsilon_0} \sum_\alpha \sum_\beta \sum_\theta \mathcal{E}_\theta \mathcal{E}_\alpha \mathcal{E}_\beta \, Im\,(\wp_{n\theta\alpha\beta} e^{i\varphi_{n\theta\alpha\beta}}) \quad (5.11.17)$$

$$\Omega_n - (\nu_n + \dot{\phi}_n) = \frac{\nu}{2\epsilon_0 \mathcal{E}_n} \sum_\alpha \sum_\beta \sum_\theta \mathcal{E}_\theta \mathcal{E}_\alpha \mathcal{E}_\beta \, Re\,(\wp_{n\theta\alpha\beta} e^{i\varphi_{n\theta\alpha\beta}}) \quad (5.11.18)$$

The amplitude function has been presented in detail in the book by Sargent, Scully and Lamb. They describe the saturation terms including the cross-saturation and the cross pushing effect. These effects have been shown in terms of separate coefficients in the case of two-mode operation.

Problems

5.1 Consider a cavity of length 10 cm. Find the frequency difference of the successive modes in GHz. Also find the wavelength difference between the modes in nm if the mode number is 10,000.

5.2 Consider a single mode in a cavity, where the electric field amplitude is time dependent and the phase is time independent.

$$\mathcal{E}(z,t) = \mathcal{E}_0(t)\exp\{-i(\nu_0 t + \phi_0 - K_0 z)\}$$

The induced polarization P_0 has the same time and z dependence. As assumed in the text the conductivity is small and the electric field and polarization are slowly varying in time. Obtain the Self-consistency equations.

5.3 Obtain the self-consistency equations in terms of the real and imaginary parts of susceptibility $\chi_n{}'$ and $\chi_n{}''$ for a single mode n as given by Eq. (5.8.11) and (5.8.12).

5.4 Write down rate equation for the electric field amplitude and also the oscillation frequency of a single mode by considering only the linear term in the polarization. Obtain the time dependent electric field amplitude and explain the gain or loss in the medium. Give physical interpretation of the frequency in this case.

Further reading

1. M. Sergeant III, M.O. Scully and W.E Lamb Jr., Laser Physics, Addison-Wesley, London, 1974.
2. S. C. Rand, Non-linear and Quantum Optics, Oxford, New York, 2010.
3. P. Meystre and M. Sargent III, Elements of Quantum Optics, Springer, Berlin, 2007.
4. M. O. Scully and M.S. Zubairy, Quantum Optics, Cambridge University Press, Cambridge, 1997.

Chapter 6
Quantum Theory of Radiation

6.1 Quantum nature of radiation

We have, so far, described the interaction of classical electromagnetic radiation field with the quantum mechanical matter, like a single atom. This theory can explain most of the features in quantum optics and laser spectroscopy. However, the semiclassical model cannot explain some of the very important features like the spontaneous emission, Lamb shift, resonance fluorescence and certain non-classical states of light. Electromagnetic radiation played an important role in the development of modern physics in the early twentieth century. The Michelson- Morley experiment on light led to the formulation of special theory of relativity, while Rayleigh-Jeans formulation on Black Body radiation led Planck to propose the quantum of action that finally marked the beginning of quantum mechanics. Planck thought of quantizing the energy of the oscillators in a cavity, but did not think of quantization of radiation. There was no concept of photon. The photon was first conceived by Einstein in his interpretation of photo-electric effect. The interaction of atom with radiation field was explained by Einstein in terms of stimulated emission, absorption and spontaneous emission. After the development of quantum mechanics, perturbation theory could be used to describe the Einstein B-coefficients. But the A-coefficients could not be derived. It could be explained by Einstein's thermodynamic argument only. This needed the quantization of the radiation field.

The quantum theory of radiation was formulated by Dirac in 1927, when he combined the wave nature of light with the particle behavior and could explain the interference phenomena. He also explained the atomic behavior of absorbing a discrete quantum of radiation for excitation from one level to another. In the quantum theory of radiation, we associate each quantum of radiation with a quantized simple harmonic oscillator and hence assign photon as a quantum of electromagnetic radiation. This theory establishes the fluctuation of zero-point energy. This is the so-called vacuum fluctuation. Like many other phenomena in quantum mechanics, it also has no classical analogue.

6.2 Maxwell's equations of classical electrodynamics in free space.

In classical electrodynamics Maxwell's equations in free space are

$$\nabla \,\mathrm{X}\, \boldsymbol{\mathcal{H}} = \frac{\partial \boldsymbol{D}}{\partial t}, \tag{6.2.1}$$

$$\nabla \,\mathrm{X}\, \mathcal{E} = -\frac{\partial \boldsymbol{B}}{\partial t}, \tag{6.2.2}$$

$$\nabla . \mathcal{E} = 0, \tag{6.2.3}$$

$$\nabla . \boldsymbol{B} = 0. \tag{6.2.4}$$

Current and free charges are assumed to be zero. If ϵ_0 and μ_0 are permittivity and permeability of the medium in free space we have

$$\boldsymbol{D} = \epsilon_0 \mathcal{E}, \tag{6.2.5}$$

$$\boldsymbol{B} = \mu_0 \boldsymbol{\mathcal{H}}. \tag{6.2.6}$$

In terms of the vector potential, $\boldsymbol{A}$

$$\boldsymbol{B} = \nabla \text{X} \boldsymbol{A}. \tag{6.2.7}$$

From Eq. (6.2.2) and (6.2.7), we get

$$\nabla \text{X} \left(\mathcal{E} + \frac{\partial A}{\partial t} \right) + 0. \tag{6.2.8}$$

Hence, we may write

$$\mathcal{E} + \frac{\partial A}{\partial t} = -\boldsymbol{\nabla}\varphi, \tag{6.2.9}$$

where φ is known as the scalar potential. Substituting in Eq. (6.2.3), we get

$$\nabla^2 \varphi + \frac{\partial}{\partial t} (\nabla . A) = 0. \tag{6.2.10}$$

From Eq. (6.2.7), we find that the vector $\boldsymbol{B}$ remains unchanged if we add gradient of any scalar to the vector potential, i.e.

$$\boldsymbol{A}' = \boldsymbol{A} + \boldsymbol{\nabla}\chi, \tag{6.2.11}$$

where χ is any scalar function. We may choose χ such that

$$\boldsymbol{\nabla}.\boldsymbol{A} = 0. \tag{6.2.12}$$

This is known as Coulomb Gauge. The scalar potential now satisfies

$$\nabla^2 \varphi = 0.$$

We may assume $\varphi = 0$ as one of the solutions. Then we get

$$\mathcal{E} = -\frac{\partial A}{\partial t}. \tag{6.2.13}$$

From Eq. (6.2.1), we can write

$$\nabla X \nabla X \boldsymbol{A} = -\mu_0 \epsilon_0 \frac{\partial^2 \boldsymbol{A}}{\partial t^2}. \tag{6.2.14}$$

Since

$$\nabla X \nabla X A = \nabla (\nabla . A) - \nabla^2 A,$$

we have

$$\nabla^2 \boldsymbol{A} = \mu_0 \epsilon_0 \frac{\partial^2 \boldsymbol{A}}{\partial t^2} = \frac{1}{c^2} \frac{\partial^2 \boldsymbol{A}}{\partial t^2}. \tag{6.2.15}$$

where, $c = (\mu_0 \epsilon_0)^{-1/2}$ is the velocity of light in free space. This is the wave equation for electromagnetic wave in free space.

6.3 Solution of the wave equation

Since the equation (6.2.15) has a space derivative and a time derivative on two sides of the equation we can separate the space and time variables

$$A(r,t) = A(r)T(t). \tag{6.3.1}$$

Hence,

$$T(t)\nabla^2 \mathbf{A}(\mathbf{r}) = \frac{1}{c^2}\mathbf{A}(\mathbf{r})\frac{\partial^2 \mathrm{T}}{\partial t^2}. \tag{6.3.2}$$

In Cartesian coordinates

$$\frac{c^2}{A_x}\nabla^2 A_x = \frac{1}{T}\frac{\partial^2 \mathrm{T}}{\partial t^2} = -\nu^2. \tag{6.3.3}$$

We assume that the two terms, which are separately functions of space and time, to be equal to a constant. So we get two equations

$$\frac{\partial^2 \mathrm{T}}{\partial t^2} + \nu^2 T = 0 \tag{6.3.4}$$

and

$$\nabla^2 A_x + \frac{\nu^2}{c^2}A_x = 0. \tag{6.3.5}$$

Similar equations may be written for the other components. We define the wave propagation constant k given by

$$k^2 = \frac{\nu^2}{c^2}.$$

The solutions for the time and space dependent parts are

$$T(t) = T(0)\, e^{-i\nu t} \tag{6.3.6}$$

and

$$A_x(\vec{r}) = A_x(0)\, e^{i\vec{k}.\vec{r}}. \tag{6.3.7}$$

$T(0)$ is the value of T at time t=0 and $A_x(0)$ is the value of A_x at $\vec{r} = 0$. Hence the vector potential is

$$\boldsymbol{A}(\vec{r}) = \boldsymbol{A}(0)\, e^{i\vec{k}.\vec{r}}. \tag{6.3.8}$$

From Eq. (6.2.12), we get

$$\boldsymbol{k}.\boldsymbol{A}(0) = 0. \tag{6.3.9}$$

Thus the vector $\boldsymbol{A}(0)$ is perpendicular to the propagation vector $\boldsymbol{k}$, this indicates the transverse nature of the electromagnetic radiation. The direction of the vector $\boldsymbol{A}(0)$ is the direction of polarization of the electromagnetic wave.

6.4 Confinement of radiation in a cavity

We consider the radiation to be confined in a finite cavity. If the radiation is confined in a cubical cavity (Fig. 6.1) of volume $V = L^3$, the periodic boundary condition yields

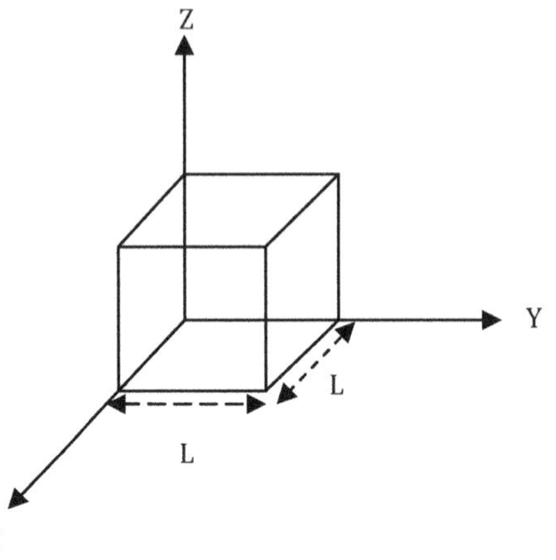

Fig 6.1 A cubical cavity of length L and volume $V = L^3$

$$\boldsymbol{A}\,(x = 0, y, z) = \boldsymbol{A}\,(x = L, y, z), \tag{6.4.1}$$

and similar relations are obtained for the y and z directions. This leads to

$$1 = e^{ik_x L} = e^{ik_y L} = e^{ik_z L}.$$

Hence

$$k_x = \frac{2\pi\vartheta_x}{L}, \tag{6.4.2}$$

$$k_y = \frac{2\pi\vartheta_y}{L}, \tag{6.4.3}$$

$$k_z = \frac{2\pi\vartheta_z}{L}, \tag{6.4.4}$$

$$\vartheta_x\,, \vartheta_y\,, \vartheta_z = 0, \pm 1, \pm 2, \pm 3, \dots \dots \dots \dots \tag{6.4.5}$$

Hence the complete solution of the vector potential is

$$\boldsymbol{A}\,(\vec{r}, t) = \sum_\zeta [\,T_\zeta\,(t)\,\boldsymbol{A}_\zeta(\vec{r}\,) + \;T^*{}_\zeta\,(t)\,\boldsymbol{A}^*{}_\zeta(\vec{r})]. \tag{6.4.6}$$

We can write

$$\boldsymbol{A}_\zeta(\vec{r}) = \boldsymbol{e}_\zeta e^{i\vec{k}_\zeta . \vec{r}} \tag{6.4.7}$$

and

$$T_\zeta\,(t) = \left|T_\zeta\right| e^{-i\nu_\zeta t}, \tag{6.4.8}$$

ζ stands for a triplet set of integer values including zero for $\vartheta_x\,,\vartheta_y\,,\vartheta_z$. The direction of polarization is given by the direction of the unit vector $\boldsymbol{e}_\zeta$. $\left|T_\zeta\right|$ corresponds to the magnitude of the vector potential for the particular set of ζ values. The second term on the right hand side of Eq. (6.4.6) is the complex conjugate of the first term, so that $\vec{A}\,(\,\vec{r},t)$ is real.

The orthogonal nature of the vector $\boldsymbol{A}_\zeta(\vec{r})$ can be expressed as

$$\iiint \boldsymbol{A}_\zeta.\boldsymbol{A}^*{}_\mu\; d\tau \;=\; \iiint \boldsymbol{A}_\zeta.\boldsymbol{A}_{-\mu} d\tau = \mathrm{V}\;\delta_{\zeta,\mu}\,. \tag{6.4.9}$$

The integration is over the volume of the cavity. In terms of the vector potential, we can write the electric and magnetic fields as

$$\boldsymbol{\mathcal{E}} = -\frac{\partial A}{\partial t} = \textstyle\sum_\zeta \boldsymbol{\mathcal{E}}_\zeta\,, \tag{6.4.10}$$

$$\boldsymbol{\mathcal{H}} = \frac{1}{\mu_0}\nabla \times \boldsymbol{A} \;=\; \textstyle\sum_\zeta \boldsymbol{\mathcal{H}}_\zeta\,, \tag{6.4.11}$$

where,

$$\boldsymbol{\mathcal{E}}_\zeta = i\nu_\zeta\,[T_\zeta\,(t)\,\vec{\boldsymbol{A}}_\zeta(\vec{r}) - T^*{}_\zeta\,(t)\,\vec{\boldsymbol{A}}^*{}_\zeta(\vec{r})], \tag{6.4.12}$$

$$\boldsymbol{\mathcal{H}}_\zeta = \frac{i}{\mu_0}\,\vec{k}_\zeta \times [\,T_\zeta\,(t)\,\vec{\boldsymbol{A}}_\zeta(\vec{r}) - T^*{}_\zeta\,(t)\,\vec{\boldsymbol{A}}^*{}_\zeta(\vec{r})]. \tag{6.4.13}$$

Hence the electric and magnetic fields that represent the time and space variation of the vector potential can be written as the sum of the quantities corresponding to the different sets of values of ζ. These different values of ζ represent different normal modes of the oscillatory field. This is a quantization in discrete numbers for the modes possible inside a cavity. This is a completely classical picture analogous to the modes in a transverse vibration of strings or the discrete frequencies in an open or closed tube filled with air.

The total energy of the radiation field in the cavity is

$$H = \frac{1}{2}\iiint (\epsilon_0\,\mathcal{E}.\mathcal{E} + \mu_0 \boldsymbol{\mathcal{H}}.\boldsymbol{\mathcal{H}}\,)\,d\tau\;. \tag{6.4.14}$$

$$\iiint\;\mathcal{E}.\mathcal{E}\,d\tau \;=\; -\textstyle\sum_{\zeta,\mu}\nu_\zeta\;\nu_\mu\,[\,T_\zeta T_\mu \iiint \boldsymbol{A}_\zeta.\boldsymbol{A}_\mu d\tau - T_\zeta T_\mu{}^* \iiint \boldsymbol{A}_\zeta.\boldsymbol{A}_\mu{}^*\,d\tau - T_\zeta{}^* T_\mu \iiint \boldsymbol{A}_\zeta{}^*.\boldsymbol{A}_\mu d\tau$$
$$+\,T_\zeta{}^* T_\mu{}^* \iiint \boldsymbol{A}_\zeta{}^*.\boldsymbol{A}_\mu{}^* d\tau]$$

$$= -V\;\textstyle\sum_{\zeta,\mu}\nu_\zeta\;\nu_\mu\,[\,T_\zeta T_\mu \delta_{\zeta,-\mu}\text{-}T_\zeta T_\mu{}^*\delta_{\zeta,\mu} - T_\zeta{}^* T_\mu \delta_{\zeta,\mu} + T_\zeta{}^* T_\mu{}^*\delta_{\zeta,-\mu}].$$

Hence,

$$\iiint\;\mathcal{E}.\mathcal{E}\,d\tau = -V\;\textstyle\sum_\zeta \nu_\zeta^2\,[T_\zeta T_{-\zeta} - 2T_\zeta T_\zeta{}^* + T_\zeta{}^* T_{-\zeta}{}^*]. \tag{6.4.15}$$

Similarly, for the magnetic field part of the Hamiltonian, we get

$$\iiint \boldsymbol{\mathcal{H}}\,.\boldsymbol{\mathcal{H}} d\tau = -\frac{1}{\mu_0^2}\textstyle\sum_{\zeta,\mu}\iiint [\,\vec{k}_\zeta \times \{\,T_\zeta\,(t)\,\vec{\boldsymbol{A}}_\zeta(\vec{r}) - T^*{}_\zeta(t)\,\vec{\boldsymbol{A}}^*{}_\zeta(\vec{r})\}]\,.\,[\,\vec{k}_\mu \times \{\,T_\mu(t)\,\vec{\boldsymbol{A}}_\mu(\vec{r}) - T^*{}_\mu\,(t)\,\vec{\boldsymbol{A}}^*{}_\mu(\vec{r})\}]d\tau$$

$$= \frac{V}{\mu_0{}^2 c^2} \sum_\zeta \nu_\zeta^2 \left[T_\zeta T_{-\zeta} + 2T_\zeta T_\zeta{}^* + T_\zeta{}^* T_{-\zeta}{}^*\right]. \quad (6.4.16)$$

Thus, the total Hamiltonian is

$$H = -\frac{1}{2}\epsilon_0 V \sum_\zeta \nu_\zeta^2 [T_\zeta T_{-\zeta} - 2T_\zeta T_\zeta{}^* + T_\zeta{}^* T_{-\zeta}{}^*] + \frac{1}{2}\epsilon_0 V \sum_\zeta \nu_\zeta^2 [\, T_\zeta T_{-\zeta} + 2T_\zeta T_\zeta{}^* + T_\zeta{}^* T_{-\zeta}{}^*]$$

$$= 2\epsilon_0 V \sum_\zeta \nu_\zeta^2 \, T_\zeta T_\zeta{}^*. \quad (6.4.17)$$

Thus the total classical Hamiltonian of the electromagnetic field in a cubical cavity can be written as the sum of the energies of the individual modes within the cavity.

6.5 Quantization of the radiation field

The energy of the radiation field of the classical wave has been expressed as the sum of the energies of individual modes in the cavity. We introduce the time-dependent variables

$$q_\zeta\,(t) = (\epsilon_0 V)^{1/2}\,[\,T_\zeta(t) + T_\zeta^*(t)]\,, \quad (6.5.1)$$

$$p_\zeta\,(t) = -i(\epsilon_0 V \nu_\zeta^2)^{1/2}\,[\,T_\zeta(t) - T_\zeta^*(t). \quad (6.5.2)$$

We can write

$$T_\zeta\,(t) = (4\epsilon_0 V\,\nu_\zeta^2)^{-1/2}\left[\,\nu_\zeta q_\zeta\,(t) + ip_\zeta(t)\right], \quad (6.5.3)$$

$$T_\zeta^*(t) = (4\epsilon_0 V\,\nu_\zeta^2)^{-1/2}\,[\,\nu_\zeta q_\zeta\,(t) - ip_\zeta(t)]\,. \quad (6.5.4)$$

Hence

$$(\,p_\zeta^2 + \nu_\zeta^2 q_\zeta^2\,) = 4\epsilon_0 V \nu_\zeta^2 T_\zeta T_\zeta^*. \quad (6.5.5)$$

From Eq. (6.4.17), we get

$$H = \frac{1}{2}\sum_\zeta(\,p_\zeta^2 + \nu_\zeta^2 q_\zeta^2\,) = \sum_\zeta H_\zeta, \quad (6.5.6)$$

where,

$$H_\zeta = \frac{1}{2}(\,p_\zeta^2 + \nu_\zeta^2 q_\zeta^2\,). \quad (6.5.7)$$

Thus the total classical energy of the electromagnetic field can be written as the sum of energies of individual linear harmonic oscillators of mass unity and frequency ν_ζ, if the coordinate and momentum are described by the conjugate variables q_ζ and p_ζ. This is only an analogy and there is no real coordinate or momentum in the usual sense of the term. In analogy with the mechanical case, we can consider the q_ζ and p_ζ as real linear operators that satisfy the commutation relations

$$[\,q_\zeta\,,p_\zeta\,] = q_\zeta\,p_\zeta - p_\zeta\,q_\zeta = i\hbar\,,$$

$$[\,q_\zeta\,,p_\mu\,] = 0, \qquad (\zeta \neq \mu)$$

$$[q_\zeta, q_\mu] = [p_\zeta, p_\mu] = 0$$

or $$[q_\zeta, p_\mu] = i\hbar\, \delta_{\zeta\mu}\,. \tag{6.5.8}$$

Next we introduce the dimensionless operators

$$a_\zeta(t) = (2\hbar\nu_\zeta)^{-1/2}\, [\nu_\zeta q_\zeta\,(t) + \, ip_\zeta(t)\,]\,, \tag{6.5.9}$$

$$a_\zeta^\dagger(t) = (2\hbar\nu_\zeta)^{-1/2} [\nu_\zeta q_\zeta\,(t) - \, ip_\zeta(t)\,]. \tag{6.5.10}$$

This leads to

$$[a_\zeta, a_\zeta^\dagger] = 1,$$

a_ζ and $a_\zeta^\dagger$ are proportional to T_ζ and T_ζ^*, hence they have similar time dependence.

$$a_\zeta\,(t) = \,|a_\zeta|\, e^{-i\nu_\zeta t}\,, \tag{6.5.11}$$

$$a_\zeta^\dagger(t) = \,|a_\zeta^\dagger|\, e^{i\nu_\zeta t}. \tag{6.5.12}$$

From equations (6.5.9) and (6.5.10) we can invert the relations to obtain

$$q_\zeta(t) = (\frac{\hbar}{2\nu_\zeta})^{\frac{1}{2}}\, [\, a_\zeta^\dagger(t) + \, a_\zeta\,(t)], \tag{6.5.13}$$

$$p_\zeta(t) = i(\frac{\hbar\nu_\zeta}{2})^{\frac{1}{2}}\, [\, a_\zeta^\dagger(t) - \, a_\zeta\,(t)]. \tag{6.5.14}$$

Hence, from Eq. (6.5.7)

$$H = \, \textstyle\sum_\zeta \hbar\, \nu_\zeta \left(a_\zeta^\dagger\, a_\zeta + \frac{1}{2}\right). \tag{6.5.15}$$

We can define the Number operator

$$N_\zeta = \, a_\zeta^\dagger\, a_\zeta \tag{6.5.16}$$

and then the Hamiltonian is

$$H = \, \textstyle\sum_\zeta \hbar\, \nu_\zeta \left(N_\zeta + \frac{1}{2}\right). \tag{6.5.17}$$

This Hamiltonian is identical to the sum of the Hamiltonians of a large number of Simple Harmonic Oscillators.

6.6 Single Mode Radiation Field

We can quantize a single mode of the radiation field in the same process as we quantize a single simple harmonic oscillator. In terms of the creation and annihilation operators the single mode electromagnetic field Hamiltonian is

$$H = \hbar\, \nu(\, a^\dagger a + \tfrac{1}{2})\,. \tag{6.6.1}$$

For a single mode we drop the suffixes for different modes.

The eigenvalue equation for the single mode can then be written as

$$H\,|n\rangle = \hbar\,\nu\left(n+\tfrac{1}{2}\right)|n\rangle\,,\quad \mathrm{n} = 0, 1, 2... \tag{6.6.2}$$

From the theory of harmonic oscillator, the eigenvalues of the number operator are discrete integers.

$$N|n\rangle = n|n\rangle, \tag{6.6.3}$$

where, n can be considered as the number of quanta in the eigenstate $|n\rangle$ of the number operator N.

From Eqs (6.5.3) and (6.5.9), we get

$$T_{\zeta}\,(t) = \left(\frac{\hbar}{2\epsilon_0 V\,\nu_{\zeta}}\right)^{\frac{1}{2}} a_{\zeta}(t)\,. \tag{6.6.4}$$

Similarly, we can derive $T_{\zeta}^{*}(t)$. Hence, the electric field for a single mode can be written from Eqs. (6.4.7), (6.4.12) and (6.6.4) as

$$\mathcal{E}\,(\vec{r},t) = i\left(\frac{\hbar\nu}{2\epsilon_0 V}\right)^{\frac{1}{2}}[a(t)e^{i\vec{k}.\vec{r}} - a^{\dagger}(t)e^{-i\vec{k}.\vec{r}}]\,\boldsymbol{e}\,. \tag{6.6.5}$$

This is the electric field operator in the Heisenberg representation. In the number state $|n\rangle$

$$\langle n|\mathcal{E}|n\rangle = 0. \tag{6.6.6}$$

Thus the expectation value of the electric field operator is zero. But, the expectation value $\langle\mathcal{E}^2\rangle$ is not zero.

$$\langle n|\mathcal{E}^2|n\rangle =-\frac{\hbar\nu}{2\epsilon_0 V}\langle n|\;(a(t)e^{ik.r} - a^{\dagger}(t)e^{-ik.r})^2|n\rangle$$

$$= \frac{\hbar\nu}{2\epsilon_0 V}\langle n|\,(aa^{\dagger} + a^{\dagger}a\,|n\rangle = \frac{\hbar\nu}{\epsilon_0 V}(\,n+\tfrac{1}{2}). \tag{6.6.7}$$

Thus there are fluctuations in the field about its zero ensemble average. It should be noted that the average indicated here is quantum mechanical average of measurements taken over a large number of identically prepared systems and it should not be confused with the time average. It does not mean that the electric field does not exist inside the cavity.

From Eq. (6.6.1) and Eq. (6.6.7) we note that the energy per unit volume associated with the electric field part of the Hamiltonian (Eq. (6.4.14)) is equal to the half of the energy eigenvalue for the mode.

The eigenstate of the mode in a generalized form can be written as

$$|\psi\rangle = \textstyle\sum_n c_n\,|n\rangle \tag{6.6.7}$$

The coefficient c_n indicates the probability that the mode has n quanta.

6.7 The Photon concept

It is known from standard quantum mechanics that the operators a and $a^\dagger$ have the property of annihilation and creation operators in the number space $|n\rangle$.

$$a|n\rangle = \sqrt{n}\,|n-1\rangle \tag{6.7.1}$$

$$a^\dagger|n\rangle = \sqrt{n+1}\,|n+1\rangle \tag{6.7.2}$$

Thus a has the effect of annihilating a quantum and $a^\dagger$ creates one quantum. From the energy eigenvalue, it is clear that even for zero quantum, the energy is $\hbar\nu/2$. Each additional quantum has energy of $\hbar\nu$. This quantum is associated with a particle called photon for a single mode electromagnetic field. The word photon is often used even without the concept of quantization. It is an important concept in laser physics and this interpretation should be very precise. But in a general sense, in the case of wave packets the photon comes from the localization of wave packet in the Fourier space. In such cases, photons are multimode objects as the Fourier analysis needs a superposition of modes. The photon number states are widely used in multimode fields.

For the ground state, we have

$$a|0\rangle = 0 \tag{6.7.3}$$

and

$$H\,|0\rangle = \hbar\nu\left(a^\dagger a + \tfrac{1}{2}\right)|0\rangle = \tfrac{1}{2}\hbar\nu|0\rangle. \tag{6.7.4}$$

Thus $\frac{1}{2}\hbar\nu$ is the energy eigenvalue of the ground state. Using the commutation relation

$$\left[H, a^\dagger\right] = \hbar\nu a^\dagger, \tag{6.7.5}$$

we get

$$Ha^\dagger|0\rangle = \left(a^\dagger H + \hbar\nu a^\dagger\right)|0\rangle = \hbar\nu\tfrac{3}{2}a^\dagger|0\rangle.$$

In this process, it can be shown that

$$H(a^\dagger)^n|0\rangle = \hbar\nu\,(n+1/2)(a^\dagger)^n|0\rangle \tag{6.7.6}$$

From Eq. (6.6.2), it is found that $(a^\dagger)^n|0\rangle$ is an eigenstate of the Hamiltonian with the eigenvalue $\hbar\nu\,(n+\frac{1}{2})$. Hence,

$$|n\rangle = N_n(a^\dagger)^n|0\rangle, \tag{6.7.7}$$

where N_n is the normalization constant. To find the normalization constant we note that

$$a^\dagger|0\rangle = |1\rangle, \tag{6.7.8}$$

$$a^\dagger|1\rangle = \sqrt{2}|2\rangle \tag{6.7.9}$$

and so on. By substituting, we can write

$$(a^\dagger)^n|0\rangle = \sqrt{1.2.3 \ldots (n+1)}\ |n\rangle. \tag{6.7.10}$$

Thus the normalized eigenvector for the photon number state is

$$|n\rangle = \frac{1}{\sqrt{n!}}(a^\dagger)^n|0\rangle. \tag{6.7.11}$$

We can obtain the coordinate representation of $|n\rangle$ as

$$\psi_n\,(q) = \langle q|n\rangle. \tag{6.7.12}$$

From expression for the annihilation operator (Eq. 6.5.9) we can write

$$(\nu q + ip)|0\rangle = 0 \tag{6.7.13}$$

or

$$(\nu q + ip)\psi_0\,(q) = 0\,, \tag{6.7.14}$$

$$\hbar\frac{d}{dq}\psi_0\,(q) = -\nu q\psi_0\,(q)\,. \tag{6.7.15}$$

Hence, the solution is

$$\psi_0\,(q) = N\exp(-\frac{\nu q^2}{2\hbar}). \tag{6.7.16}$$

$\psi_0\,(q)$ may be normalized to obtain the constant N and we get the ground state wave function as

$$\psi_0\,(q) = (\frac{\nu}{\pi\hbar})^{1/4}\exp(-\frac{\nu q^2}{2\hbar})\,. \tag{6.7.17}$$

The wave function for the n-th state may then be written as

$$\psi_n\,(q) = \frac{1}{\sqrt{n!}}(a^\dagger)^n\psi_0\,(q)$$

$$= \frac{1}{\sqrt{n!}}\{(2\hbar\nu)^{-1/2}\,[\nu q\ -\ \hbar\,\frac{d}{dq}]\}^n\ \psi_0\,(q). \tag{6.7.18}$$

The right hand side is the n-th order Hermite polynomial.

6.8 Multimode Radiation Field

In a multimode radiation field, the number state can be written as

$$|n_1, n_2, \ldots\ldots n_s \ldots\rangle = |n_1\rangle|n_2\rangle| \ldots\ldots\ldots |n_s\rangle \ldots \tag{6.8.1}$$

This is a quantum state for the multimode field with n_1 photons in the mode 1, n_2 photons in mode 2 and so on.

For the multimode Hamiltonian the eigenvalue equation is

$$H|n_1, n_2, \ldots n_s \ldots\rangle = \Sigma_s(n_s + \tfrac{1}{2})\ \hbar\nu_s|n_1, n_2, \ldots\ldots n_s \ldots\rangle. \tag{6.8.2}$$

The eigenvalue is a sum of eigenvalues of the modes that are not interacting with each other and are independent modes in the cavity. The annihilation and creation operators are also specific and act on that mode only,

$$a_s|n_1, n_2, \dots \dots n_s \dots\rangle = \sqrt{n_s}|n_1, n_2, \dots \dots n_s - 1 \dots\rangle, \quad (6.8.3)$$

$$a_s^{\dagger}|n_1, n_2, \dots \dots n_s \dots\rangle = \sqrt{n_s + 1}|n_1, n_2, \dots \dots n_s + 1 \dots\rangle. \quad (6.8.4)$$

Hence, the creation or annihilation operator corresponding to a mode creates or destroys a photon of that particular mode, when all other modes remain unchanged. Similarly, for the expectation value of the electric field operator we have

$$\langle n_1, n_2, n_3, \dots \dots n_s \dots | \mathcal{E}_s|n_1, n_2, n_3, \dots \dots n_s \dots\rangle = \quad \langle n_1|n_1\rangle\langle n_2|n_2\rangle \dots \dots \langle n_s| \mathcal{E}_s |n_s\rangle \dots . = 0. \quad (6.8.5)$$

$$\langle n_1, n_2, n_3, \dots \dots n_s \dots | \mathcal{E}_s^2 |n_1, n_2, n_3, \dots \dots n_s \dots\rangle$$

$$= \frac{\hbar\nu_s}{\epsilon_0 V}\left(n_s + \frac{1}{2}\right). \quad (6.8.6)$$

The uncertainty in the Electric field as operator is

$$\Delta\mathcal{E}_s{}^2 = \langle\mathcal{E}_s^2\rangle - \langle\mathcal{E}_s\rangle^2 = \frac{\hbar\nu_s}{\epsilon_0 V}\left(n_s + \frac{1}{2}\right). \quad (6.8.7)$$

In a general case, the state vector can be written as a linear super position of the different multimode states

$$|\psi\rangle = \sum_{n_1}\sum_{n_2} \dots . \sum_{n_s} \dots C_{n_1 n_2 \dots n_s \dots}|n_1, n_2, \dots \dots n_s \dots\rangle \quad (6.8.8)$$

This state includes the state vectors of the multimode which are correlated and may result from interaction with atomic systems. The coefficients $C_{n_1 n_2 \dots n_s \dots}$ represent the probability of finding a state with n_1 photons in mode 1, n_2 photons in mode 2 and so on.

The multimode electric field operator is

$$\mathcal{E}(\vec{r}, t) = i\sum_{\varsigma}\left(\frac{\hbar\nu_{\varsigma}}{2\epsilon_0 V}\right)^{\frac{1}{2}}[a_{\varsigma}(t)e^{i\boldsymbol{k}_{\varsigma}.\boldsymbol{r}} - a^{\dagger}{}_{\varsigma}e^{-i\boldsymbol{k}_{\varsigma}.\boldsymbol{r}}]\,\boldsymbol{e}_{\varsigma}. \quad (6.8.9)$$

If the electromagnetic field propagates in the z direction the electric field operator in the x direction is

$$\mathcal{E}_x(z, t) = -\sum_{\varsigma}\left(\frac{2\hbar\nu_{\varsigma}}{\epsilon_0 V}\right)^{\frac{1}{2}}(a_{\varsigma} + a^{\dagger}{}_{\varsigma})\sin k_{\varsigma}z. \quad (6.8.10)$$

Thus, the real operator for the electric field is expressed in terms of the creation and annihilation operators.

6.9 Coherent State

A set of quantum mechanical states minimizes the Heisenberg uncertainty product $\Delta p\Delta q$, which leads to

$$\Delta p\Delta q = \hbar/2, \quad (6.9.1)$$

where,

$$\Delta p = (\langle p^2\rangle - \langle p\rangle^2)^{1/2} \quad (6.9.2)$$

and

$$\Delta q = (\langle q^2\rangle - \langle q\rangle^2)^{1/2} \tag{6.9.3}$$

are the root mean square deviations of the operators p and q. Coherent states form a special set of states that simultaneously minimizes both the uncertainties of momentum and position co-ordinates.

Coherent states are closest approach to the classical states. Since they have nearly classical character, we look for the pure states of the harmonic oscillator that have mean energy equal to the energy of the classical oscillator. According to the Ehrenfest's theorem for a harmonic oscillator, as well as that of free particle, the dynamics of the quantum wave packet follows the classical laws of physics. Thus the expectation values of the time dependent operators $q(t)$ and $p(t)$ are the classical time dependent variables $Q(t)$ and $P(t)$.

$$\langle\psi|q(t)|\psi\rangle = Q(t)\,. \tag{6.9.4}$$

$$\langle\psi|p(t)|\psi\rangle = P(t)\,. \qquad (6.9\ 5)$$

Substituting these expressions into the Hamiltonian of classical harmonic oscillator (Eq. 6.5.7) we get

$$H = \frac{1}{2}\,(\,v^2 Q^2 + \,P^2\,)$$

$$=\frac{1}{2}(v^2\langle\psi|q(t)|\psi\rangle^2 + \,\langle\psi|p(t)|\psi\rangle^2) \tag{6.9.6}$$

Substituting the expressions (Eq. 6.5.13) and (Eq. 6.5.14) for the operators q and p for a single mode, we get the classical harmonic oscillator from Eq. (6.9.6)

$$H = \,\hbar v\,\langle\psi|a|\psi\rangle\langle\psi|a^\dagger|\psi\rangle. \tag{6.9.7}$$

The energy of the corresponding quantum mechanical oscillator, excluding the zero point energy $\hbar v/2$, is

$$\langle H\rangle = \,\hbar v\,\langle\psi|a^\dagger a|\psi\rangle, \tag{6.9.8}$$

We require that the classical energy of the harmonic oscillator be equal to the quantum mechanical average for the state ψ. Hence,

$$\langle\psi|a^\dagger a|\psi\rangle = \,\langle\psi|a|\psi\rangle\langle\psi|a^\dagger|\psi\rangle\,. \tag{6.9.9}$$

Following Glauber, the states that satisfy these conditions are called Coherent states. The left hand side of the equation (6.9.9) is a first order correlation function for the operator a, where the classical ensemble average is replaced by a quantum mechanical average and the right hand side is the product of the average of the same operator and its conjugate. This average is a complex number as the operator is not hermitian. This is Glauber condition.

We can show that the coherent state $|\psi\rangle$ described by the coherence condition of Eq. (6.9.9) is an eigenstate of the annihilation operator. For this purpose, we note that the Eq. (6.9.9) can be written as

$$\langle\psi|a^\dagger a|\psi\rangle \quad =(\langle\psi|a^\dagger|\psi\rangle)^2. \tag{6.9.10}$$

From the Gram-Schmidt orthogonalization procedure, starting with $|\psi\rangle$, we can construct a complete orthonormal set consisting of states $|\psi\rangle$ and an infinite complimentary basis set of vectors $\{|\Phi\rangle\}$. In terms of the identity operator we can write

$$I = |\psi\rangle\langle\psi| + \sum_{\Phi} |\Phi\rangle\langle\Phi|. \quad (6.9.11)$$

Introducing the identity operator we can write

$$\langle\psi|a^{\dagger}a|\psi\rangle = \langle\psi|a^{\dagger}|\psi\rangle\langle\psi|a|\psi\rangle + \langle\psi|a^{\dagger}\langle \sum_{\Phi} |\Phi\rangle\langle\Phi|a|\psi\rangle$$

$$= |\langle\psi|a^{\dagger}|\psi\rangle|^{2} + \sum_{\Phi}|\langle\Phi|a|\psi\rangle|^{2}. \quad (6.9.12)$$

Comparing with (6.9.10), we get

$$\langle\Phi|a|\psi\rangle = 0, \quad (6.9.13)$$

for all Φ, since each term in the summation over Φ is positive and should be separately zero. Hence $a|\psi\rangle$ must be orthogonal to any $|\Phi\rangle$ or it should be proportional to $|\psi\rangle$. This state satisfying the condition of Eq. (6.9.9) is a coherent state. Following standard notation for coherent states, we write $|\psi\rangle$ as $|\alpha\rangle$. Hence we can write

$$a|\alpha\rangle = \alpha|\alpha\rangle. \quad (6.9.14)$$

The constant α is the eigenvalue of the annihilation operator. Since a is not a hermitian operator, the eigenvalue is a complex number. We can now find the eigenvalue and the eigenstate. The state $|\alpha\rangle$ can be written as a complete set in terms of the photon number states $|n\rangle$ as

$$|\alpha\rangle = \sum_{n} C_{n} |n\rangle, \quad (6.9.15)$$

$$a|\alpha\rangle = \alpha|\alpha\rangle = \alpha \sum_{n} C_{n} |n\rangle, \quad (6.9.16)$$

$$a|\alpha\rangle = \sum_{n} C_{n}\, a|n\rangle = \sum_{n} C_{n} \sqrt{n}|n-1\rangle. \quad (6.9.17)$$

Equating the coefficients of $|n\rangle$ from the two equations (6.9.16) and (6.9.17) we can write

$$C_1 = \alpha C_0,$$

$$C_2 = \frac{\alpha C_1}{\sqrt{2}} = \frac{\alpha^2}{\sqrt{2!}} C_0.$$

Similarly,

$$C_n = \frac{\alpha^n}{\sqrt{n!}} C_0. \quad (6.9.18)$$

Hence

$$|\alpha\rangle = C_0 \sum_{n} \frac{\alpha^n}{\sqrt{n!}} |n\rangle. \quad (6.9.19)$$

By normalization, we get

$$\langle\alpha|\alpha\rangle = |C_0|^2 \sum_{m}\sum_{n} \frac{\alpha^n(\alpha^*)^m}{\sqrt{n!}\sqrt{m!}} \langle m|n\rangle$$

$$= |C_0|^2 \sum_{m}\sum_{n} \frac{\alpha^n(\alpha^*)^m}{\sqrt{n!}\sqrt{m!}} \delta_{mn} = |C_0|^2 e^{|\alpha|^2}. \quad (6.9.20)$$

Hence

$$|\alpha\rangle = e^{-|\alpha|^2/2}\sum_n \frac{\alpha^n}{\sqrt{n!}}|n\rangle \tag{6.9.21}$$

excluding an arbitrary phase factor. The eigenvalue α can take any complex value.

The probability of finding n photons in the coherent state $|\alpha\rangle$ is

$$P_\alpha(n) = \langle n|\alpha\rangle^2 = e^{-|\alpha|^2}\frac{\alpha^{2n}}{n!} \tag{6.9.22}$$

This is called Poisson distribution that is centered at $n = |\alpha|^2$ and has width $|\alpha|$ as shown in Fig. 6.2.

In summary we can state that coherent state has the following characteristics.

This satisfies the condition that the uncertainty product is a minimum.

It satisfies the Glauber condition.

It is an eigenstate of the annihilation operator.

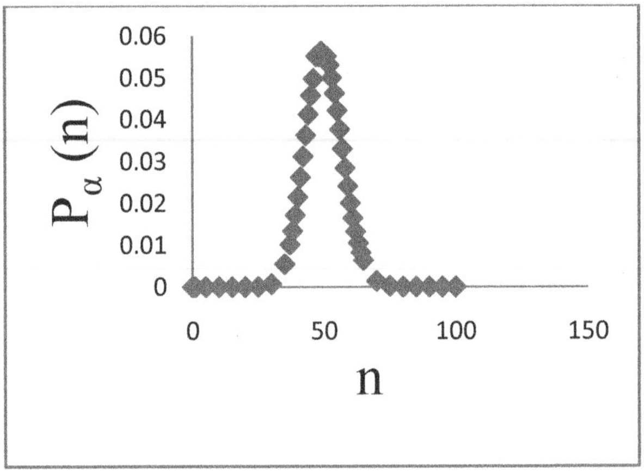

Fig 6.2 Poisson's distribution for coherent state of Eq.(6.9.22) centered about n=50

Substituting $|n\rangle$ in terms of $|0\rangle$ from Eq. (6.7.8) in Eq. (6.7.20)

$$|\alpha\rangle = e^{-|\alpha|^2/2}\sum_n \frac{(\alpha a^\dagger)^n}{n!}|0\rangle$$

$$= e^{-|\alpha|^2/2}e^{\alpha a^\dagger}|0\rangle$$

$$= e^{-|\alpha|^2/2}e^{\alpha a^\dagger}e^{-\alpha^* a}|0\rangle. \tag{6.9.23}$$

The last term in the equation can be introduced, since $a|0\rangle = 0$. We can use Baker-Hausdorff relation

$$e^{A+B} = e^A e^B e^{-[A,B]/2}. \tag{6.9.24}$$

Since $[\alpha a^\dagger, \alpha^* a] = -|\alpha|^2$, eq. (6.9.23) can be written as

$$|\alpha\rangle = D(\alpha)|0\rangle, \tag{6.9.25}$$

where,

$$D(\alpha) = e^{\alpha a^{\dagger} - \alpha^{*} a} \qquad (6.9.26)$$

is called the displacement operator that displaces the vacuum state to the coherent state.

It can be verified that the coherent states are not orthogonal to each other.

$$\langle \alpha | \beta \rangle = \exp\left(-\frac{1}{2}\left[\ |\alpha|^2 + |\beta|^2 - 2\,\alpha^{*}\beta\ \right]\right)$$

$$|\langle \alpha | \beta \rangle|^2 = \exp(-|\alpha - \beta|^2\,) \qquad (6.9.27)$$

This is not zero, when $\alpha \neq \beta$. Hence the states with different eigenvalues are not orthogonal. If α differs considerably from β, the states are closer to the orthogonality condition.

Problems

6.1 For a single mode radiation field obtain an expression for the electric field uncertainty $\Delta\mathcal{E}$.

6.2 For the single mode radiation field calculate the magnetic field uncertainty $\Delta\mathcal{H}$. Hence obtain the uncertainty relation.

6.3 For the coherent state $|\alpha\rangle$ show that
$\int d^2\alpha|\alpha\rangle\langle\alpha| = \pi\sum_n|n\rangle\langle n| = \pi.$

6.4 For the coherent state $|\alpha\rangle$ show that the minimum uncertainty product is given by
$\Delta p\Delta q = \hbar/2.$

6.5 Show that the displacement operator satisfies the following relation

$$D^\dagger(\alpha) = D(-\alpha) = [D(\alpha)]^{-1}.$$

6.6 Show that
$D^{-1}(\alpha)aD(\alpha) = \alpha + a,$
$D^{-1}(\alpha)a^\dagger D(\alpha) = \alpha^* + a^\dagger.$

6.7 Obtain the coordinate representation $\psi_\alpha(q)$ of the coherent state $|\alpha\rangle$.

6.8 An electric field operator in terms of the two oppositely running travelling waves can be written as
$\mathcal{E}(t) = \mathcal{E}_0[(a_1 - a_2^\dagger)e^{ikz} + (a_2 - a_1^\dagger)e^{-ikz} +]e^{-ivt},$
write down the same field in terms of the annihilation operators of the standing wave
$a_c = (a_1 + a_2)/\sqrt{2}$ and $a_s = (a_1 - a_2)/\sqrt{2}$
$\mathcal{E}_0$ is the electric field per photon for both travelling waves. The standing wave electric field per photon is given be $\mathcal{E}_s = \sqrt{2}\mathcal{E}_0$.

6.9 Show that the number operator does not commute with the electric field operator and the commutation relation is

$$[n, \mathcal{E}_x] = \mathcal{E}_0 \sin kz\,(a^\dagger - a),$$

where the single mode electric field is obtained from Eq. (6.8.10). Hence find the uncertainty relation between the number operator and the electric field operator.

6.10 The quadrature operators are defined as

$$X = \frac{1}{2}(a + a^\dagger) \quad Y = \frac{1}{2i}(a - a^\dagger).$$

Show that

$$[X, Y] = \frac{i}{2} \quad \text{and} \quad \langle\Delta X\rangle\langle\Delta Y\rangle \geq \frac{1}{4}.$$

Further reading

1. P. Meystre and M. Sargent III, Elements of Quantum Optics, Springer, Berlin, 2007.
2. M. O. Scully and M.S. Zubairy, Quantum Optics, Cambridge University Press, Cambridge, 1997.
3. S. C. Rand, Non-linear and Quantum Optics, Oxford, New York, 2010.
4. K. Thyagarajan and A. Ghatak, Lasers: Fundamentals and Applications, Springer, London, 2010.
5. C. C. Gerry and P. L. Knight, Introductory Quantum Optics, Cambridge University Press, Cambridge, 2005.

Chapter 7
Quantum Theory of Atom Field Interaction

7.1 Atom-Field Hamiltonian in terms of Pauli operators

We consider a two-level atom (Fig. 7.1) having an upper energy level b and a lower energy level a with energy values $E_a = \hbar\omega_a$ and $E_b = \hbar\omega_b$, the resonance frequency is defined as $\omega = \omega_b - \omega_a$. The eigenfunctions φ_a and φ_b of the two-level atom can be written as

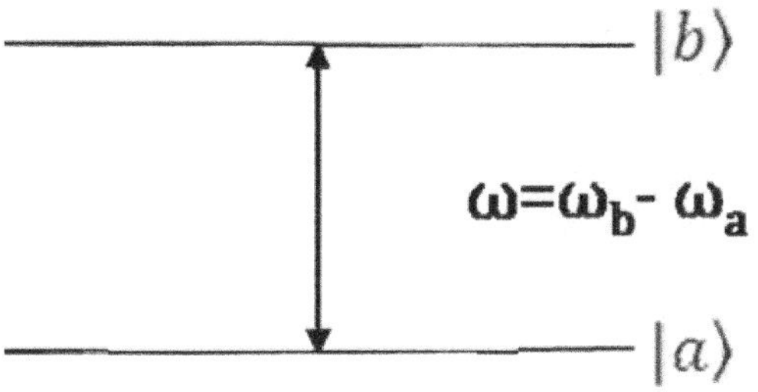

Fig. 7.1 Two-level atom with resonance frequency ω.

$$\varphi_a = \binom{0}{1} \quad \text{and} \quad \varphi_b = \binom{1}{0}. \tag{7.1.1}$$

The atomic wave function is written as

$$\psi = C_a\varphi_a + C_b\varphi_b. \tag{7.1.2}$$

So the state vector is

$$\psi = \binom{C_b}{C_a}.$$

We can define the non-hermitian spin-flip operators in terms of the Pauli spin matrices,

$$\sigma_x = \begin{pmatrix} 0 & 1 \\ 1 & 0 \end{pmatrix}, \; \sigma_y = \begin{pmatrix} 0 & -i \\ i & 0 \end{pmatrix}, \; \sigma_z = \begin{pmatrix} 1 & 0 \\ 0 & -1 \end{pmatrix} \tag{7.1.3}$$

as

$$\sigma_+ = \frac{1}{2}(\sigma_x + i\,\sigma_y) = \begin{pmatrix} 0 & 1 \\ 0 & 0 \end{pmatrix}, \tag{7.1.4}$$

$$\sigma_- = \frac{1}{2}(\sigma_x - i\,\sigma_y) = \begin{pmatrix} 0 & 0 \\ 1 & 0 \end{pmatrix}. \tag{7.1.5}$$

Hence,

$$\sigma_+\varphi_a = \varphi_b \quad \text{and} \quad \sigma_-\varphi_b = \varphi_a \tag{7.1.6}$$

σ_+ and σ_- are the raising and lowering operators of the atomic states. If the energy is considered to be equal to zero at the level halfway between the two levels, then

$$E_a = -\frac{1}{2}\hbar\omega \quad \text{and} \quad E_b = +\frac{1}{2}\hbar\omega.$$

So the atomic Hamiltonian can be written as

$$H_{atom} = \frac{1}{2}\hbar\omega\, \sigma_z. \tag{7.1.7}$$

The electric field operator for the single mode (Eq.6.8.10) can be written as

$$\mathcal{E} = \mathcal{E}_0\,(a + a^\dagger)\,\sin kz\,, \tag{7.1.8}$$

$$\mathcal{E}_0 = \left(\frac{2\hbar\nu}{\epsilon_0 V}\right)^{\frac{1}{2}}. \tag{7.1.9}$$

The interaction energy of the atom and the quantized field can be written, in the dipolar interaction case, in the same way as in the semi classical theory (Eq. 2.2.5). Using RWA we can write the interaction energy as

$$V = \hbar g(\sigma_+ + \sigma_-)(a + a^\dagger), \tag{7.1.10}$$

where,

$$g = \frac{\mu\mathcal{E}_0}{2\hbar}\sin kz \tag{7.1.11}$$

is the Rabi frequency term described earlier in the semi classical case. The factor ½ in Eq. 7.1.8 arises from RWA. The dipole moment component for the two-level atom at rest is taken in the direction of the electric field. The axis of quantization for the atom is taken such that g is real.

Hence the total atom field Hamiltonian is

$$H = \frac{1}{2}\hbar\omega\,\sigma_z + \hbar\nu\left(a^\dagger a + \frac{1}{2}\right) + \hbar g(\sigma_+ + \sigma_-)(a + a^\dagger). \tag{7.1.12}$$

The interaction energy expressed by the last term, as given above, has four terms signifying the absorption or emission process. These also include terms that are energy non-conserving and may be omitted in the subsequent discussion.

7.2 Absorption and emission process

The term $\sigma_+ a$ describes a process in which the atom is raised from the lower level a to the upper level b and one photon is annihilated or absorbed by the atom (Fig. 7.2). This is an energy conserving process. The term $\sigma_- a^\dagger$ indicates the downward transition of the atom from the upper level b to the lower level a and the emission

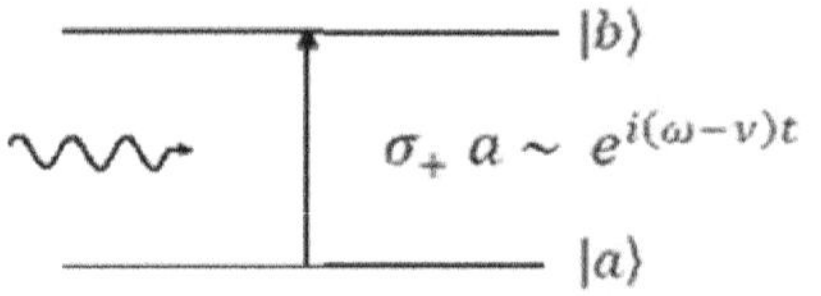

Fig. 7.2 The absorption of a photon incident on a two level atom.

of a photon. This is also an energy conserving process and represents stimulated emission (Fig. 7.3).

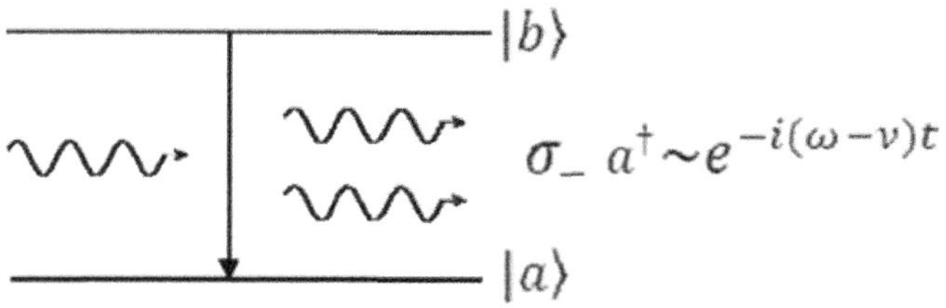

Fig. 7.3 The emission of a photon as the atom moves to the lower level.

The term $\sigma_+ a^\dagger$ correspondsto an upward transition of the atom with simultaneous emission of a photon. This is not allowed from the principle of conservation of energy (Fig. 7.4).

The term $\sigma_- a$ corresponds to a downward transition of the atom and simultaneously absorption of a photon. This process does not conserve energy (Fig. 7.5).

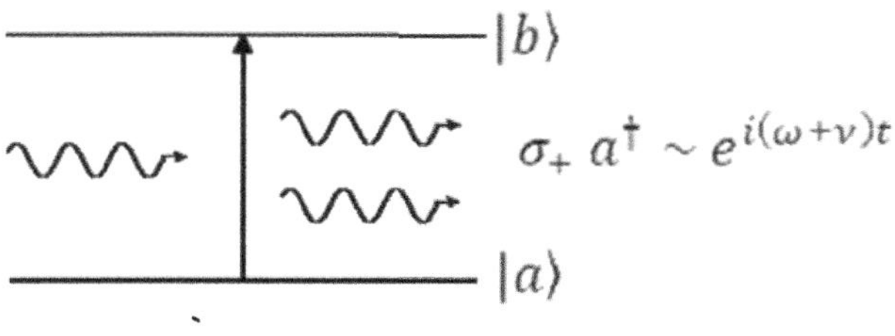

Fig. 7.4 An energy non-conserving process with emission of a photon and upward transition of the atom .

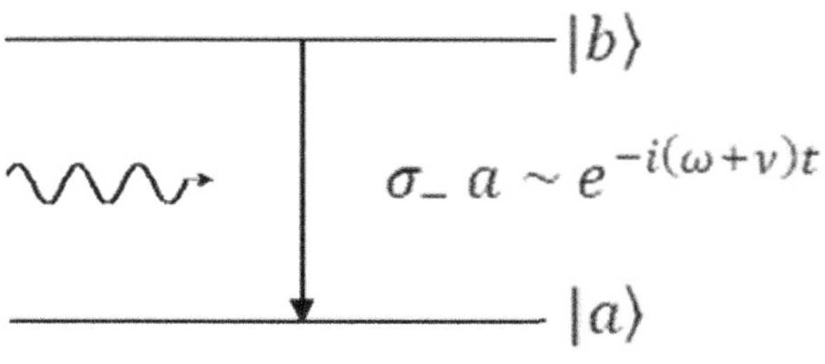

Fig. 7.5 An energy non-conserving process with the absorption of a photon and downward transition of the atom.

As we shall explain now, we shall retain only the energy conserving terms in the interaction energy and omit the other two terms. Considering the time dependence of the operators in the Heisenberg picture, the time dependence of spin flip operators are

$$\sigma_\pm (t) = \sigma_\pm (0)\, e^{\pm i\omega t} . \tag{7.2.1}$$

And as given in Chapter 6, we have the time dependence of a and $a^\dagger$ as

$$a\,(t) = a(o)e^{-i\nu t} \quad \text{and} \quad a^\dagger\,(t) = a^\dagger(0)\, e^{i\nu t}. \tag{7.2.2}$$

Hence,

$$\sigma_+(t)\, a\,(t) = \sigma_+(0)\, a\,(0)\;\; e^{i(\omega-\nu)t}. \tag{7.2.3}$$

Similarly, we can write

$$\sigma_-(t)\, a^\dagger\,(t) = \sigma_-(0)\, a^\dagger\,(0)\;\; e^{-i(\omega-\nu)t}, \tag{7.2.4}$$

$$\sigma_+(t)\, a^\dagger(t) = \sigma_+(0)\, a^\dagger\,(0)\;\; e^{i(\omega+\nu)t}, \tag{7.2.5}$$

$$\sigma_-(t)\, a\,(t) = \sigma_-(0)\, a\,(0)\;\; e^{-i(\omega+\nu)t}. \tag{7.2.6}$$

Eq 7.2.3 and 7.2.4 show that these terms vary with time at low frequency near resonance, where detuning is small; whereas, the terms in Eq. 7.2.5 and 7.2.6 are rapidly varying anti-resonance terms. They average to zero in time after a few optical periods and hence physically they have no contributions. It may be noted that in semi classical perturbation treatment, it is shown that time integration will lead to larger denominators for such cases. This corresponds to the same physics as we have in case of RWA. For these reasons, in our subsequent treatment we shall omit these terms and consider the interaction Hamiltonian as

$$H_{int} \;=\; \hbar g(\sigma_+ a +\; \sigma_- a^\dagger\,) \tag{7.2.7}$$

7.3 Dressed States

The unperturbed atom-field state for the first two terms of the Hamiltonian of Eq. (7.1.12) is

$$|an\rangle = |a\rangle|n\rangle\,. \tag{7.3.1}$$

Where, the field has n photons and the atom is in state a. The unperturbed Hamiltonian H_0

$$H_0 \;=\; \frac{1}{2}\hbar\omega\;\sigma_z + \hbar\nu\;\left(\,a^\dagger a + \frac{1}{2}\right) \tag{7.3.2}$$

satisfies the eigenvalue equations

$$H_0|an\rangle = \{\,-\frac{1}{2}\hbar\omega +\;\hbar\nu\;\left(\,n + \frac{1}{2}\right)\}|an\rangle, \tag{7.3.3}$$

$$H_0|bn\rangle = \{\frac{1}{2}\hbar\omega +\;\hbar\nu\;\left(\,n + \frac{1}{2}\right)\}\;|bn\rangle. \tag{7.3.4}$$

The interaction term $\sigma_+ a$ of the Hamiltonian couples the states $|an+1\rangle$ and $|bn\rangle$

$$\sigma_+ a|an+1\rangle = \sqrt{n+1}\;|bn\rangle. \tag{7.3.5}$$

Similarly,

$$\sigma_- a^\dagger|bn\rangle = \sqrt{n+1}\;|an+1\rangle. \tag{7.3.6}$$

The lower state has one more photon than the upper state, so that there is conservation for the closed two-level coupled system under the action of a monochromatic or single mode radiation field at resonance frequency.

Caution: There should be no confusion between the symbols for annihilation operator a and the notation for the atomic state a that does not correspond to any physical quantity.

The other term like $\sigma_+\; a^\dagger$ or $\sigma_- a$ would connect states like

$|an-1\rangle$ and $|bn\rangle$ or $|bn+1\rangle$ and $|an\rangle$. Such processes are not physically allowed and are dropped by RWA. Hence we consider the manifold of states $|bn\rangle$ and $|an+1\rangle$ as forming a basis set

$$\begin{pmatrix} |bn\rangle \\ |an+1\rangle \end{pmatrix}.$$

and in this basis, the total atom field Hamiltonian from Eq. (7.2.7) and (7.3.1-6) is

$$H = \hbar\nu \begin{pmatrix} n & 0 \\ 0 & n+1 \end{pmatrix} + \frac{1}{2}\hbar\omega \begin{pmatrix} 1 & 0 \\ 0 & -1 \end{pmatrix} + \hbar g \begin{pmatrix} 0 & \sqrt{n+1} \\ \sqrt{n+1} & 0 \end{pmatrix}, \tag{7.3.7}$$

where, we have dropped the zero point energy term. Introducing the detuning term $\delta = \omega - \nu$ we can rewrite the Hamiltonian as

$$H = \hbar\nu\,(n+1/2) \begin{pmatrix} 1 & 0 \\ 0 & 1 \end{pmatrix} + \hbar/2 \begin{pmatrix} \delta & 2g\sqrt{n+1} \\ 2g\sqrt{n+1} & -\delta \end{pmatrix}. \tag{7.3.8}$$

The factor ½ in the first term arises from mathematical manipulation and has nothing to do with the zero point energy which is not included. When $n = 0$, the interaction connects the ground state $|a1\rangle$ with $|b0\rangle$ and there is absorption. Similarly, there can be emission of a photon and the transition takes place from $|b0\rangle$ to $|a1\rangle$. This shows the result of interaction with one manifold of states $\{|bn\rangle, |an+1\rangle\}$. The total Hamiltonian acts on all other similar manifolds with different photon numbers. The interaction matrix is written in terms of Rabi frequency, this is also true for the classical Hamiltonian. But in this case, the Rabi term is multiplied by a factor $n+1$. This is the contribution of quantum nature of radiation. The diagonalization of the second term of the matrix leads to the atom-field coupled states. The eigenvalue equation can be written as

$$\begin{pmatrix} \delta & 2g\sqrt{n+1} \\ 2g\sqrt{n+1} & -\delta \end{pmatrix} \begin{pmatrix} |2n\rangle \\ |1n\rangle \end{pmatrix} = \Lambda_n \begin{pmatrix} |2n\rangle \\ |1n\rangle \end{pmatrix}.$$

This leads to

$$\Lambda_n^2 = \delta^2 + 4g^2\,(n+1),$$

$$\Lambda_n = \pm\sqrt{\delta^2 + 4g^2\,(n+1)} = \pm\Omega_n, \tag{7.3.9}$$

$$E_{2n} = \hbar\nu\left(n+\frac{1}{2}\right) - \frac{1}{2}\hbar\Omega_n, \tag{7.3.10}$$

$$E_{1n} = \hbar\nu\left(n+\frac{1}{2}\right) + \frac{1}{2}\hbar\Omega_n, \tag{7.3.11}$$

Ω_n is the Rabi flopping frequency. At Resonance ($\omega = \nu$), it is given by $2g\sqrt{n+1}$.

The energy eigenvectors can be calculated as

$$|2n\rangle = \cos\theta_n\,|bn\rangle - \sin\theta_n\,|an+1\rangle, \tag{7.3.12}$$

$$|1n\rangle = \sin\theta_n|bn\rangle + \cos\theta_n\,|an+1\rangle. \tag{7.3.13}$$

Where it can easily be shown that

$$\cos\theta_n = \frac{\Omega_n - \delta}{\sqrt{(\Omega_n-\delta)^2 + 4g^2\,(n+1)}}, \tag{7.3.14}$$

$$\sin\theta_n = \frac{2g\sqrt{n+1}}{\sqrt{(\Omega_n-\delta)^2 + 4g^2\,(n+1)}}. \tag{7.3.15}$$

At resonance, $\delta = 0$ and $\Omega_n = 2g\sqrt{n+1}$, hence, $\cos\theta_n = \sin\theta_n = \frac{1}{\sqrt{2}}$ and $\theta_n = \frac{\pi}{4}$. The states $|2n\rangle$ and $|1n\rangle$ are the dressed states and can be obtained through matrix of rotation by angle θ_n from the unperturbed basis states.

$$T_n = \begin{pmatrix} \cos\theta_n & -\sin\theta_n \\ \sin\theta_n & \cos\theta_n \end{pmatrix}, \tag{7.3.16}$$

From Eq. (7.3.12) and (7.3.13), we can write

$$T_n \begin{pmatrix} |bn\rangle \\ |an+1\rangle \end{pmatrix} = \begin{pmatrix} |2n\rangle \\ |1n\rangle \end{pmatrix}. \tag{7.3.17}$$

It should be noted that even when there is no photon or $n = 0$, $\sin\theta_n \neq 0$ and $\cos\theta_n \neq 1$, this means that there is still dressing of the states. These dressed states in the case of no photons are

$$|2\,0\rangle = \frac{1}{\sqrt{(\Omega_0-\delta)^2 + 4g^2}}[(\Omega_0 - \delta)|b0\rangle - 2g\,|a1\rangle] \tag{7.3.18}$$

$$|1\,0\rangle = \frac{1}{\sqrt{(\Omega_0-\delta)^2 + 4g^2}}[2g|b0\rangle + (\Omega_0 - \delta)\,|a1\rangle] \tag{7.3.19}$$

The eigenvalues of the dressed states may be plotted against the atomic emission frequency ω (Fig. 7.6). The unperturbed eigenvalues E_{bn+1} and E_{an} are represented by the straight lines (dotted) and they cross each other at $\omega = \nu$. The dressed state eigenvalues are shown as the curved lines. They repel each other at the resonance frequency. This repulsion is called anti-crossing. The minimum separation occurs at resonance and then it is equal to $2\hbar g\sqrt{n+1}$. The dressed atom energy includes the interaction energy, whereas the bare atom energy does not include the same.

At resonance, the dressed states are

$$|2n\rangle = \frac{1}{\sqrt{2}}\,[\,|bn\rangle - |an+1\rangle] \tag{7.3.20}$$

$$|1n\rangle = \frac{1}{\sqrt{2}}\,[\,|bn\rangle + |an+1\rangle] \tag{7.3.21}$$

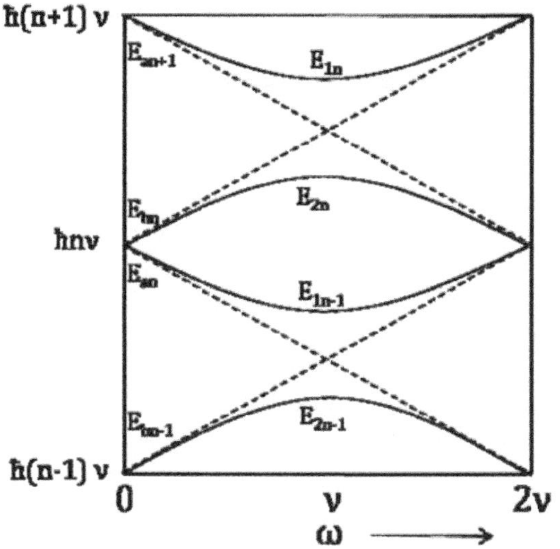

Fig. 7.6 Energy level diagram for dressed and bare atom states. The dotted straight lines show the unperurbed energy of the bare states. The curved lines represent the energy of the dressed states.

The energy eigenvalues are

$$E_{2n} = \hbar\nu\left(n+\frac{1}{2}\right) - \hbar g\sqrt{n+1}, \tag{7.3.22}$$

$$E_{1n} = \hbar\nu\left(n+\frac{1}{2}\right) + \hbar g\sqrt{n+1}. \tag{7.3.23}$$

In the case of very large detuning, the RWA breaks down and the diagonalization process is invalid.

In the Schrödinger picture a general state vector can be represented in terms of the bare atom state vectors as

$$|\psi\rangle = \Sigma_n[\,C_{an+1}(t)|an+1\rangle + C_{bn}(t)\,|bn\rangle\,]. \tag{7.3.24}$$

The same vector in terms of the dressed states can be written as

$$|\psi\rangle = \Sigma_n[\,C_{2n}(t)|2n\rangle + C_{1n}(t)\,|1n\rangle\,]. \tag{7.3.25}$$

The probability amplitudes are related by

$$\begin{pmatrix} C_{2n}(t) \\ C_{1n}(t) \end{pmatrix} = T_n \begin{pmatrix} C_{bn}(t) \\ C_{an+1}(t) \end{pmatrix}. \tag{7.3.26}$$

Entangled states

The coupling of two or more quantum mechanical states representing two different quantum mechanical systems leads to a new class of quantum mechanical states called entangled states. In this case, a two level atom, coupled to a single mode quantized electromagnetic field, results in the dressed state. The Dressed State is an example of Entangled States. This state exhibits a quantum correlation between the two entities, atom and field, so that one particular system cannot be separately prescribed. As an example, if the system is in a state $|1n\rangle$ we cannot say whether the atom is in state a or b, but if in any measurement we find the system in the atomic state b, then we are sure that the field has $n+1$ photons.

If two systems are entangled by interaction, they will remain entangled in space even when they are physically separated, if there is no dissipation or decoherence. Entagled states cannot be factored into a simple product of the eigenstates of the two interacting subsystems like atom and field. It is obvious from above that neither of $|2n\rangle$ or $|1n\rangle$ can be factored as $|a\rangle|n\rangle$ or $|b\rangle|n+1\rangle$

$$|\psi\rangle_{entangled} \neq |\psi_{atom}\rangle\, |\psi_{field}\rangle$$

The entangled states form the basis of quantum information science, quantum teleportation or quantum computing to be discussed in Chapter 10.

7.4 Rate equation treatment of atom field interaction

We consider a two-level atom interacting with a single mode radiation of frequency ν. In absence of the field the initial atom-field state vector is

$$|\psi_{at}\rangle = C_a|a\rangle + C_b|b\rangle. \tag{7.4.1}$$

The state vector for the electromagnetic field is

$$|\psi_f\rangle = \sum_n C_n|n\rangle\,. \tag{7.4.2}$$

The ket vectors $|a\rangle, |b\rangle$ are the atomic eigenvectors and $|n\rangle$ denotes the photon number state as in the previous section. The total unperturbed atom field state vector is

$$|\psi_{at-f}\rangle = \sum_n [\, C_{an}|a\rangle|n\rangle + C_{bn}|b\rangle|n\rangle]. \tag{7.4.3}$$

The atom is assumed to interact with n number of photons. If we assume that at time t=0, the atom is in upper state b and the field has n photons then the state at t=0 is

$$|\psi_{at-f}\rangle_{t=0} = |b\rangle|n\rangle \tag{7.4.4}$$

When the field is switched on, the atom starts interacting with the field and there can be transition to the lower level with a change in the number of photons. Hence at a later time t, the atom field state is

$$|\psi_{at-f}\rangle_t = C_{bn}|bn\rangle| + C_{an+1}|an+1\rangle. \tag{7.4.5}$$

This means that the atom has made a transition to the level a and emitted a photon. The total atom field Hamiltonian as described in Section 7.1 (Eq. 7.1.12) is

$$H = \hbar\begin{pmatrix}\omega_b & 0\\ 0 & \omega_a\end{pmatrix} + \hbar\nu\left(a^\dagger a + \frac{1}{2}\right) + \hbar g(\sigma_+ + \sigma_-)(a + a^\dagger). \tag{7.4.6}$$

The notations have same meaning as in Section 7.1 The perturbation in the time dependent form may be obtained by similarity transformation with the corresponding unperturbed Hamiltonian. In the interaction picture the interaction energy can be written as

$$\hbar g \exp[\, iH_{atom}t/\hbar](\sigma_+ + \sigma_-)\ \exp[-\frac{iH_{atom}t}{\hbar}]\ \times\ \exp[\frac{iH_{field}t}{\hbar}]\,(a + a^\dagger)\ \exp[-\frac{iH_{field}t}{\hbar}]$$

where H_{atom} corresponds to the first term of Eq. 7.4.6 and H_{field} is the second term of Eq. 7.4.6. By algebraic manipulation we can obtain the time dependence in the interaction Hamiltonian as

$$H_{int} = \hbar g(\sigma_+ e^{i\omega t} + \sigma_- e^{-i\omega t})(ae^{-ivt} + a^\dagger e^{ivt}) \tag{7.4.7}$$

Using RWA we can omit the energy non-conserving terms, so

$$H_{int} = \hbar g(\sigma_+ a e^{i(\omega-v)t} + \sigma_- a^\dagger e^{-i(\omega-v)t}). \tag{7.4.8}$$

This is the Hamiltonian in the Interaction picture and may be compared with the Hamiltonian of Eq. (7.2.7). The equation of motion in the interaction picture is

$$|\dot{\psi}\rangle = -\frac{i}{\hbar} H_{int} |\psi\rangle. \tag{7.4.9}$$

Using the state vector of Eq.(7.4.5) and the interaction Hamiltonian of Eq. (7.4.8), we get

$$\dot{C}_{bn}|bn\rangle + \dot{C}_{an+1}|an+1\rangle = -ig[\sigma_+ a e^{i(\omega-v)t} + \text{adj}] \times [C_{bn}|bn\rangle| + C_{an+1}|an+1\rangle].$$

Multiplying by $|bn\rangle$ we have

$$\dot{C}_{bn} = -ig\sqrt{n+1}e^{i(\omega-v)t}C_{an+1}. \tag{7.4.10}$$

Similarly,

$$\dot{C}_{an+1} = -ig\sqrt{n+1}e^{-i(\omega-v)t}C_{bn}. \tag{7.4.11}$$

If the atom is initially in the lower state at t=0,

$$C_{an+1}(0) = 1, \ C_{bn}(0) = 0 .$$

In the first order perturbation theory, we assume the condition to remain valid for all time t, Solution of equation (7.4.11) as

$$C_{bn} = -ig\sqrt{n+1}\int_0^t e^{i(\omega-v)t}\,dt$$

$$= -ig\sqrt{n+1}\frac{e^{i(\omega-v)t}-1}{i(\omega-v)}. \tag{7.4.12}$$

So we can get the probability of the atom to be in state b and emit one photon as

$$|C_{bn}(t)|^2 = g^2t^2(n+1)\frac{sin^2(\omega-v)t/2}{[(\omega-v)t/2]^2}. \tag{7.4.13}$$

If the initial condition is chosen such that the atom is in the upper state with n photons

$$C_{an+1}(0) = 0, \ C_{bn}(0) = 1.$$

Following the same procedure, we can write

$$|C_{an+1}(t)|^2 = g^2t^2(n+1)\frac{sin^2(\omega-v)t/2}{[(\omega-v)t/2]^2}. \tag{7.4.14}$$

This is the probability of emission in the case of n photons in the upper state. It is proportional to $n+1$. In the case of zero photons, the probability is non-zero, in the resonance case it is

$$|C_{a1}(t)|^2 = g^2t^2. \tag{7.4.15}$$

This is spontaneous emission in the case of a single mode, since it occurs even in the case of zero photons. This is not a derivation of spontaneous emission, but it indicates the existence of spontaneous emission. In the next section we shall consider the theory of spontaneous emission.

The term n in the emission probability corresponds to the stimulated emission.

The equation of motion may be solved exactly in case of resonance $\omega = \nu$. Taking the derivative of rate equation (7.4.10) we get

$$\ddot{C}_{bn}(t) = -ig\sqrt{n+1}\dot{C}_{an+1}(t)$$

$$=-g^2(n+1)C_{bn}(t). \tag{7.4.16}$$

The general solution of the equation is

$$C_{bn}(t) = A\ \sin g\sqrt{n+1}\,\mathrm{t} + B\cos g\sqrt{n+1}\ t \tag{7.4.17}$$

From equation (7.4.10), with $\omega = \nu$ we can write

$$C_{an+1} = i(g\sqrt{n+1})^{-1}\dot{C}_{bn}$$

$$= iA\cos g\sqrt{n+1}\ t \text{ - } iB\sin g\sqrt{n+1}\,t.$$

For a resonant atom initially in the upper level, we can have $C_{an+1}(0) = 0,\ C_{bn}\,(0) = 1$ and hence $B = 1$ and $A = 0$, so

$$C_{bn}(t) = \cos g\sqrt{n+1}\ t, \tag{7.4.18}$$

$$C_{an+1} = \text{ - } i\sin g\sqrt{n+1}\,t. \tag{7.4.19}$$

Hence the probability for the states is

$$|C_{bn}(t)|^2 = \cos^2 g\sqrt{n+1}\,t\,, \tag{7.4.20}$$

$$|C_{an+1}(t)|^2 = \sin^2 g\sqrt{n+1}\,t\,. \tag{7.4.21}$$

This shows Rabi flopping between the levels. In this case of emission, the probability of emission is given by $|C_{an+1}(t)|^2$. For the case of absorption we have the initial condition, $C_{an+1}(0) = 1,\ C_{bn}\,(0) = 0$, this leads to

$$B = 0 \ \text{ and } A = 1,$$

and hence, we have

$$|C_{bn}(t)|^2 = \sin^2 g\sqrt{n+1}\,t\,, \tag{7.4.22}$$

$$|C_{an+1}(t)|^2 = \cos^2 g\sqrt{n+1}\,t\,. \tag{7.4.23}$$

The probabilities of emission and absorption oscillate in time with the frequency$g\sqrt{n+1}$. This may be compared with the classical case of strong field. In the case of strong field the number of photons is very large and the factor $n+1$ may be approximated by n and we have classical correspondence. In a weak field, this is not true and so the classical correspondence and hence semi classical theory is invalid.

7.5 Theory of spontaneous emission: Wigner-Weisskopf model

We noted in the previous section that in the case of resonance, an atom can oscillate between the two levels in the presence of a single mode electromagnetic field. This can happen even when there is no photon. For an atom initially in the upper state it can flip to the lower level, but without any photon it is not possible for the photon to move to the upper level. Hence this theory is not correct. The problem arises because of the assumption of a single mode. In general, there is a continuum of modes in the case of an infinitely large cavity. Hence a large number of modes with zero photon are always present in the system. We have to consider a multimode field.

The emission of a photon by an atom creates an electromagnetic field that can act with the atom itself. For a single mode field, this produces two frequencies slightly above and slightly below the resonance frequency. This is analogous to the case of two coupled oscillators. In a multimode field this produces a band of frequencies giving rise to a line width of the emitted spectrum.

We consider a two level atom interacting with a multimode field. The Hamiltonian of the system is

$$H = \sum_k \hbar\nu_k \left(a_k^\dagger a_k + \tfrac{1}{2}\right) + \hbar\begin{pmatrix}\omega_b & 0\\ 0 & \omega_a\end{pmatrix} +$$

$$\hbar\sum_k g_k\,(\sigma_+ a_k + \sigma_- a_k^\dagger). \qquad (7.5.1)$$

In the interaction picture, the interaction energy is

$$H_{int} = \sum_k \hbar g_k(\sigma_+ a_k e^{i(\omega-\nu_k)t} + \sigma_-\, a_k^\dagger e^{-i(\omega-\nu_k)t})$$

$$= \sum_k H_k.$$

The unperturbed eigenvector is

$$|\alpha,\ n_1, n_2, \dots\dots n_k \dots\dots\dots\rangle \qquad (\alpha = a, b)$$

The general state vector can be written as

$$|\psi\rangle = \sum_{\alpha=a,b}\sum_{n_1}\sum_{n_2}\dots\dots\sum_{n_k}\dots . C_{\alpha,n_1,n_2,\dots\dots n_k\dots\dots}$$

$$|\alpha,\ n_1, n_2, \dots\dots n_k \dots\dots\dots\rangle. \qquad (7.5.2)$$

This represents a state that is a mixture of states in which the atom is in one of the two states and the first mode has n_1 photons, the second mode has n_2 photons etc. The C-coefficients denote the probability of such states. Summation over each n_k runs from 0 to∞.

In the interaction picture, the equation of motion is the same as Eq. (7.4.9) with the Hamiltonian and the state vector as given above.

$$\sum_{\alpha=a,b}\sum_{n_1}\sum_{n_2}\dots\sum_{n_k}\dots . \dot{C}_{\alpha,n_1,n_2,\dots\dots n_k\dots\dots}\,|\,\alpha, n_1, n_2, \dots\dots n_k \dots\rangle$$
$$= -\frac{i}{\hbar}\sum_k H_k \sum_{\alpha=a,b}\sum_{n_1}\sum_{n_2}\dots\dots\sum_{n_k}\dots . C_{\alpha,n_1,n_2,\dots\dots n_k\dots\dots}$$

$$|\alpha,\ n_1, n_2, \dots \dots n_k \dots \rangle \tag{7.5.3}$$

Taking projection onto the state $\langle \alpha,\ n_1, n_2, \dots \dots n_k \dots |$ we get

$$\dot{C}_{\alpha,n_1,n_2,\dots\dots n_k \dots\dots} = -\frac{i}{\hbar}\sum_{\beta=a,b}\sum_{m_1}\sum_{m_2}\dots\dots\sum_{m_l}\dots\langle \alpha,\ n_1, n_2, \dots \dots n_k \dots |\sum_l H_l\,|\beta, m_1, m_2 \dots . m_l \dots\rangle C_{\beta,m_1,m_2 \dots m_l \dots} . \tag{7.5.4}$$

H_k is the single mode interaction energy that affects the k-th mode only. The matrix element may be calculated as

$$\langle\, \alpha, n_1, n_2, \dots \dots n_k \dots \sum_l H_l\,|\beta, m_1, m_2 \dots . m_l \dots\rangle = \sum_l \langle \alpha n_l|H_l|\beta m_l\rangle \langle\, n_1, n_2, .\,. n_{l-1}, n_{l+1} \dots |m_1, m_2 \dots . m_{l-1}, m_{l+1} .\,.\rangle$$

$$= \sum_l \langle \alpha n_l|H_l|\beta m_l\rangle\, \delta_{n_1 m_1}\delta_{n_2 m_2}\delta_{n_{l-1} m_{l-1}}\delta_{n_{l+1} m_{l+1}} \dots .$$

Hence the Eq. (7.5.4) becomes

$$\dot{C}_{\alpha,n_1,n_2,\dots\dots n_k \dots\dots} = -\frac{i}{\hbar}\sum_\beta \sum_l \sum_{m_l}\langle \alpha n_l|H_l|\beta m_l\rangle\, C_{\beta,n_1,n_2,\dots\dots m_l \dots\dots} . \tag{7.5.5}$$

The matrix elements of σ operators in H_k are

$$\langle \alpha|\sigma_-|\beta\rangle = \delta_{\alpha a}\delta_{\beta b}\,, \tag{7.5.6}$$

$$\langle \alpha|\sigma_+|\beta\rangle = \delta_{\alpha b}\delta_{\beta a}\,. \tag{7.5.7}$$

We can evaluate the matrix element of the Hamiltonian as

$$\langle \alpha n_l|H_l|\beta m_l\rangle$$

$$=\langle \alpha n_l|\hbar g_l(\sigma_+ a_l e^{i(\omega-\nu_l)t} + \sigma_-\, a_l^\dagger e^{-i(\omega-\nu_l)t}|\beta m_l\rangle$$

$$= \hbar g_l(\, \delta_{\alpha b}\delta_{\beta a}\delta_{n_l m_l-1}\sqrt{m_l}e^{i(\omega-\nu_l)t} +$$

$$(\, \delta_{\alpha a}\delta_{\beta b}\delta_{n_l m_l+1}\sqrt{m_l+1}\, e^{-i(\omega-\nu_l)t}). \tag{7.5.8}$$

Hence, the rate equation for the C coefficient is

$$\dot{C}_{\alpha,n_1,n_2,\dots\dots n_k \dots\dots}$$

$$=-i\sum_l g_l\,(\sqrt{n_l+1}e^{i(\omega-\nu_l)t}\ \delta_{\alpha b}C_{a,n_1,n_2,\dots\dots n_l+1\dots\dots}$$

$$+ \sqrt{n_l}e^{-i(\omega-\nu_l)t}\ \delta_{\alpha a}C_{b,n_1,n_2,\dots\dots n_l-1\dots\dots}) \tag{7.5.9}$$

We can consider an initial state, where the atom is in the upper state and all modes have zero photon or no electromagnetic field is applied, we can write

$$C_{b,0,0,0,\dots\dots 0_k\dots\dots\dots}(t=0) = 1 \tag{7.5.10}$$

or, $\quad C_{b,0}(0) = 1.$

All other C -coefficients are zero. The Hamiltonian can connect a state $|b,0\rangle$ with the states $|a,1_k\rangle$, where the atom will be in the lower state with one photon in the k-th mode and no photon in any other mode. The state vector (Eq. 7.5.2) can then be written as

$$|\psi\rangle = C_{b,0}\ |b,0\rangle + \sum_k C_{a,1_k}\ |a,1_k\rangle. \tag{7.5.11}$$

In course of time, there may be a change in the probability amplitude and the particle may be in the lower state with one photon in the k-th mode or

$$C_{a,0,0,0,........1_k......}(t) = C_{a,1_k}(t)\ . \tag{7.5.12}$$

The equation of motion for the coefficients may then be written from Eq. (7.5.9) as

$$\dot{C}_{b,0} = -i\sum_k g_k\, e^{i(\omega-\nu_k)t}C_{a,1_k}, \tag{7.5.13}$$

$$\dot{C}_{a,1_k} = -ig_k e^{-i(\omega-\nu_k)t}\ C_{b,0}. \tag{7.5.14}$$

The above two equations show that there can be emission from any of the k-th modes from the upper state, the ground state can get a photon only from one mode as it has a photon in the k-th mode. Taking the integral of Eq (7.5.4) and substituting in Eq. (7.5.13), we get

$$\dot{C}_{b,0} = -\ \sum_k g_k^2 \int_0^t d\tau\, e^{i(\omega-\nu_k)(t-\tau)}\ C_{b,0}(\tau). \tag{7.5.15}$$

In the case of very closely spaced frequencies or a continuum of modes, we can replace the summation by an integral in terms of the density of states $D(\nu)$ as

$$\Sigma \quad \rightarrow \int d\nu\, D(\nu).$$

The density of states can be obtained by assuming the radiation to be confined in a cavity of volume$V = L^3$. Along the x direction, the running modes in the cavity can have wave numbers $K_x = 2\pi\vartheta_x/L,\ \vartheta_x = 1, 2, \ldots$ and so on for y and z directions. The number of modes in interval between K_x and $K_x + dK_x$ is $d\vartheta_x = dK_x\, L/2\pi$. Considering the modes in the three dimensional cavity the number of modes is

$$d\,\vartheta = d^3K\,\left(\frac{L}{2\pi}\right)^3. \tag{7.5.16}$$

For large K the summation over K can be replaced by an integration

$$1/V\sum_K dK\,F(K) \rightarrow \frac{1}{V}\int d\,\vartheta\,F(K) = \frac{1}{(2\pi)^3}\int d^3K\,F(K).$$

Using spherical polar coordinates,

$$d^3K = dKK^2 \sin\theta\ d\theta\ d\varphi$$

and using the frequency $\nu = cK$, we have

$$\frac{1}{V}\sum_K dK\,F(K) = \frac{1}{(2\pi)^3}\int d\nu\,\frac{\nu^2}{c^3}\int_0^\pi \sin\theta d\theta\,\int_0^{2\pi} d\varphi F(\nu).$$

We have also to sum over two polarization components of the electromagnetic field.

$$\frac{1}{V}\sum_{K} dK\, F(K) = \int d\nu\, D(\nu)F(\nu)$$

where the density of states is

$$D(\nu) = \frac{V\nu^2}{\pi^2 c^3}. \tag{7.5.17}$$

The density of states and the coupling term g do not vary rapidly with ν as compared to the oscillatory term, so they can be brought out of the integral over ν. Since $\omega - \nu$ is small, when the radiation frequency is close to resonance, they can assume the same values as ω

$$\dot{C}_{b,0} = -\int d\nu g^2(\nu)\, D(\nu) \int_0^t d\tau\, e^{i(\omega-\nu)(t-\tau)}\, C_{b,0}(\tau)$$

$$= -\, g^2(\nu)\, D(\nu) \int d\nu \int_0^t d\tau\, e^{i(\omega-\nu)(t-\tau)}\, C_{b,0}(\tau). \tag{7.5.18}$$

We can assume that the term $C_{b,0}(\tau)$ varies sufficiemtly slowly so that it can be evaluated at $\tau = t$ and the time integral is

$$\int_0^t d\tau\, e^{i(\omega-\nu)(t-\tau)} = \pi\delta(\omega - \nu) - \wp\left(\frac{i}{\omega-\nu}\right), \quad , \tag{7.5.19}$$

where the second term indicates the principal value. This term leads to the level shift related to the Lamb shift. The first term leads to

$$\dot{C}_{b,0}(t) = -\frac{1}{2}\gamma C_{b,0}(t). \tag{7.5.20}$$

where, the decay term is

$$\gamma = 2\pi g^2(\nu)\, D(\nu), \tag{7.5.21}$$

$$|C_{b,0}(t)|^2 = e^{-\gamma t}. \tag{7.5.22}$$

Thus an atom in the excited state, in the absence of a field, decays exponentially with time. This is Wigner Weisskopf model for spontaneous emission.

Problems

7.1 Show that

$$[\sigma_+, \sigma_-] = \sigma_z$$
$$[\sigma_-, \sigma_z] = 2\,\sigma_-$$

7.2 If two operators A and B do not commute, show that

$$e^{\alpha A} B e^{-\alpha A} = B + \alpha[A, B] + \frac{1}{2!}\alpha^2[A, [A, B]] + \cdots$$

7.3 With the help of the result in Problem no. 2 show that

$$e^{iva^\dagger at} a e^{-iva^\dagger at} = a e^{-ivt}$$
$$e^{i\omega\sigma_z t/2} \sigma_+ e^{-i\omega\sigma_z t/2} = \sigma_+ e^{i\omega t}$$

7.4 A two-level atom is interacting with a n-photon radiation field. If the number of photons is very large and nearly equal to 10^6 and both detuing and Rabi frequency are small, of the order 10 MHz, write down the approximate form of normalized dressed states. Find the energy level splitting of the dressed states.

Further reading

1. M. Sergeant III, M.O. Scully and W.E Lamb Jr., Laser Physics, Addison-Wesley, London, 1974.
2. S. C. Rand, Non-linear and Quantum Optics, Oxford, New York, 2010.
3. P. Meystre and M. Sargent III, Elements of Quantum Optics, Springer, Berlin, 2007.
4. M. O. Scully and M.S. Zubairy, Quantum Optics, Cambridge University Press, Cambridge, 1997.
5. K. Thyagarajan and A. Ghatak, Lasers: Fundamentals and Applications, Springer, London, 2010.
6. C. C. Gerry and P. L. Knight, Introductory Quantum Optics, Cambridge University Press, Cambridge, 2005.

Chapter 8
Doppler-free spectroscopy with laser

8.1 Introduction

In Chapter 3 we discussed how linear spectroscopy involving two atomic levels shows a Gaussian profile because of random atomic velocities under thermal distribution. The Doppler broadening is usually large and half width is of the order of 200 to 300 MHz at room temperature. Atomic features like collision effect, power saturation and all other atomic phenomena are suppressed under this umbrella of Doppler profile. Homogeneous broadening caused by collision, power or transit time broadening is normally of the order of a few MHz. In Chapter 4 we found that in case of saturation absorption spectroscopy we can use counter propagating laser beams that can create holes on the population distribution curves. Two counter propagating lasers can produce two holes symmetrically placed on either side of the Gaussian peak depending on the detuning. When the detuning is zero we can obtain a Lorentzian hole at the centre representing the Lamb dip. The Lamb dip reveals some of the atomic features. In this experiment on Lamb dip, both the beams are of equal intensity. However, one can adjust the reflected counter-propagating beam to have lower strength if the mirror is not fully reflecting. In such cases one weak beam may act as probe and it will monitor the full effect of the changes created by the stronger pump beam.

Saturation spectroscopy is more conveniently performed in three level atomic systems interacting with two laser radiations. One of the radiations, called pump beam, has higher intensity and frequency on resonance or close to resonance with the transition frequency across two levels of the atom. This radiation causes saturation and produces significant change of population difference by absorption. The second laser radiation, called probe beam, is weaker and interacts with two other energy levels, one of which is common to the pump transition. It has a frequency on resonance or close to resonance of the transition frequency of the latter pair of energy levels. The radiation probes the changes caused by the pump radiation. There can be three different energy level configurations with (a, b, c) as the energy levels for such pump-probe system:- (1) Cascade-type (Fig. 8.1(a)), (2) V-type (Fig. 8.1(b)) and (3) Λ- type (Fig. 8.1 (c)). In all the configurations the level b is the common level for both pump and probe.

In this chapter we shall solve the density matrix for a three –level cascade system and obtain the probe absorption coefficient in presence of the pump beam. We shall later discuss the two other level configurations. We shall show that several non-linear features can be demonstrated and the line shape of sub-natural line width can be obtained by manipulation of the powers of the two interacting laser radiations. The interaction of the laser radiation can produce a coherent superposition of the atomic states such that atoms are trapped in one state so that the atomic system may be transparent even in the presence of radiation frequency that is on resonance. The absorption or transmission may be controlled by a laser beam.

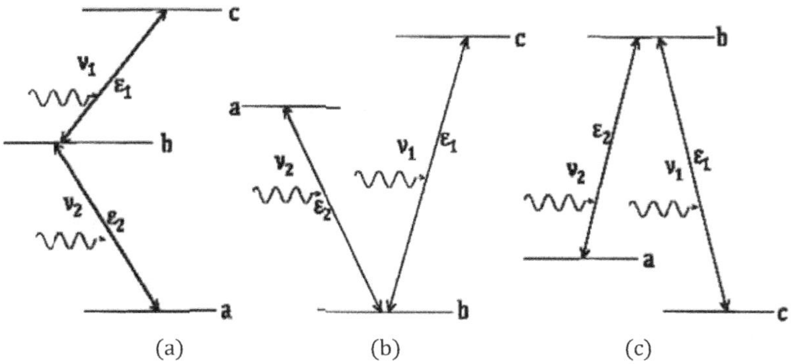

Fig. 8.1 Energy levels of a three level atom for (a) cascade, (b) V-type and (c) Λ- type.

8.2 Density matrix equations for three-level system

We consider two electromagnetic fields having electric field amplitudes $\mathcal{E}_{10}$ and $\mathcal{E}_{20}$ and frequencies ν_1 and ν_2 interacting with the atomic levels causing transitions $a \rightarrow b$ and $b \rightarrow c$ respectively (Fig. 8.1(a)).

$$\mathcal{E}_1 = \mathcal{E}_{10} \cos(\nu_1 t - k_1 z), \tag{8.2.1}$$

$$\mathcal{E}_2 = \mathcal{E}_{20} \cos(\nu_2 t - k_2 z). \tag{8.2.2}$$

The radiation frequencies are close to the transition frequencies ω_{ba} and ω_{cb} and $|\omega_{ba} - \omega_{cb}|$ is large enough so that a radiation acts upon only one transition. The transitions $a \rightarrow b$ and $b \rightarrow c$ are called the **pump** and **probe** transitions respectively The transition dipole matrix elements are μ_{ab} and μ_{bc} and the Rabi frequencies are

$$\Omega_1 = \frac{\mu_{ab}\mathcal{E}_{10}}{\hbar} \text{ and } \Omega_2 = \frac{\mu_{bc}\mathcal{E}_{20}}{\hbar}. \tag{8.2.3}$$

The rate equations for the density matrix can easily be written following the procedure described in Section 3.3

$$\dot{\rho}_{aa} = \lambda_a - \Upsilon_a \rho_{aa} + \left(-\frac{i}{\hbar} V_{ab}\rho_{ba} + C.C.\right),$$

$$= \lambda_a - \Upsilon_a \rho_{aa} - i\Omega_1 \cos(\nu_1 t - k_1 z)(\rho_{ba} - \rho_{ab}). \tag{8.2.4}$$

Similarly,

$$\dot{\rho}_{bb} = i\,\Omega_1 \cos(\nu_1 t - k_1 z)(\rho_{ba} - \rho_{ab}) - i\Omega_2 \cos(\nu_2 t + k_2 z)(\rho_{cb} - \rho_{bc}) - \Upsilon_b \rho_{bb}, \tag{8.2.5}$$

$$\dot{\rho}_{cc} = i\Omega_2 \cos(\nu_2 t - k_2 z)(\rho_{cb} - \rho_{bc}) - \Upsilon_c \rho_{cc}, \tag{8.2.6}$$

$$\dot{\rho}_{ba} = -(i\omega_{ba} + \gamma_{ba})\rho_{ba} + i\Omega_1 \cos(\nu_1 t - k_1 z)(\rho_{bb} - \rho_{aa}) - i\Omega_2 \cos(\nu_2 t - k_2 z)\rho_{ca}, \tag{8.2.7}$$

$$\dot{\rho}_{cb} = -(i\omega_{cb} + \gamma_{cb})\rho_{cb} + i\Omega_1 \cos(\nu_1 t - k_1 z)\rho_{ca} + i\Omega_2 \cos(\nu_2 t - k_2 z)(\rho_{cc} - \rho_{bb}), \tag{8.2.8}$$

$$\dot{\rho}_{ca} = -(i\omega_{ca} + \gamma_{ca})\rho_{ca} + i\Omega_1 \cos(\nu_1 t - k_1 z)\rho_{cb} - i\Omega_2 \cos(\nu_2 t - k_2 z)\rho_{ba}. \tag{8.2.9}$$

There is pumping at the rate λ_a only to the lower level. We consider rotating wave approximation and use the following transformation

$$\rho_{ba} = e^{-i(\nu_1 t - k_1 z)}\tilde{\rho}_{ba}, \tag{8.2.10}$$

$$\rho_{cb} = e^{-i(\nu_2 t - k_2 z)}\tilde{\rho}_{cb}, \tag{8.2.11}$$

$$\rho_{ca} = e^{-i[(\nu_1+\nu_2)t-(k_2+k_2)z)]}\tilde{\rho}_{ca}. \tag{8.2.12}$$

We attempt a perturbation solution by setting $\Omega_2 = 0$ to obtain a first order solution for $\tilde{\rho}_{ba}$ and ρ_{bb} and ρ_{aa}. We define the level population difference between b and a as

$$\tilde{N} = -\lambda_a/\Upsilon_a.$$

We consider the total time derivative as

$$\frac{d}{dt} = \frac{\partial}{\partial t} + v\frac{\partial}{\partial z}$$

and follow the procedure developed in section 3.5 to obtain the first order values of the matrix elements as (from Eq. 3.5.8, 3.5.15, 3.5.21 and 3.5.22)

$$\rho_{bb}{}^{(1)} = -\bar{N}(v)\frac{2I_1\eta_1\gamma_{ba}^2}{\gamma_{ba}^2 + (\Delta_1+k_1v)^2+2I_1\eta_1\gamma_{ba}^2}, \tag{8.2.13}$$

$$\rho_{aa}{}^{(1)} = -\bar{N}(v)[1 - \frac{2I_1\eta_1\gamma_{ba}^2}{\gamma_{ba}^2 + (\Delta_1+k_1v)^2+2I_1\eta_1\gamma_{ba}^2}], \tag{8.2.14}$$

$$\tilde{\rho}_{ba}{}^{(1)} = -\frac{\Omega_1}{2}\bar{N}(v)\frac{(\Delta_1+k_1v)+i\gamma_{ba}}{[(\Delta_1+k_1v)^2+\gamma_{ba}^2(1+2I_1\eta_1)]}. \tag{8.2.15}$$

The dimensionless parameters I_1 and η_1 are the same as those defined earlier in Eq. (3.5.17-18) and population is considered as a function of the velocity.

We have defined

$$\Delta_1 = \omega_{ba} - \nu_1, \tag{8.2.16}$$

$$I_1 = (\frac{\mu_{ab}\mathcal{E}_{10}}{\hbar})^2\frac{1}{2\gamma_a\gamma_b}, \tag{8.2.17}$$

$$\eta_1 = \frac{\gamma_b+\gamma_a}{2\gamma_{ba}}. \tag{8.2.18}$$

8.3 Perturbation solution

We can substitute the first order results into the remaining three equations and obtain $\tilde{\rho}_{cb}$ and $\tilde{\rho}_{ca}$. We shall retain terms upto the lowest order in $\mathcal{E}_2$. It is noticed that the term involving ρ_{cc} in the rate equation of ρ_{cb} (Eq. 8.2.8) is higher order in Ω_2 and hence can be neglected. We can obtain the two equations involving $\tilde{\rho}_{cb}$ and $\tilde{\rho}_{ca}$ as follows (using Eq. 8.2.8, 8.2.9, 8.2.10 and 8.2.11)

$$(\Delta_2 + k_2v - i\gamma_{cb})\tilde{\rho}_{cb}{}^{(2)} + \frac{\Omega_1}{2}\tilde{\rho}_{ca}{}^{(2)} = \frac{\Omega_2}{2}\rho_{bb}{}^{(1)}, \tag{8.3.1}$$

$$(\Delta_1 + \Delta_2 + (k_1+k_2)v - i\gamma_{ca})\tilde{\rho}_{ca}{}^{(2)} + \frac{\Omega_1}{2}\tilde{\rho}_{cb}{}^{(2)} = \frac{\Omega_2}{2}\tilde{\rho}_{ba}{}^{(1)}, \tag{8.3.2}$$

where, $\Delta_2 = \omega_{cb} - \nu_2$. Solving these two equations, we can obtain the second order transition matrix element

$$\tilde{\rho}_{cb}{}^{(2)} = \frac{\Omega_2}{2}\frac{(\Delta_1+\Delta_2+(k_1+k_2)v-i\gamma_{ca})\;\rho_{bb}{}^{(1)} - \frac{\Omega_1}{2}\tilde{\rho}_{ba}{}^{(1)}}{(\Delta_2+k_2v-i\gamma_{cb})(\Delta_1+\Delta_2+(k_1+k_2)v-i\gamma_{ca})-(\frac{\Omega_1}{2})^2}. \tag{8.3.3}$$

In the case of a weak field intensity we can omit the $(\frac{\Omega_1}{2})^2$ term in the denominator. In such cases, we also have very small I_1. The term arising from population of the intermediate state can be written as

$$(\tilde{\rho}_{cb}{}^{(2)})_{pop} = -\frac{\Omega_2}{2}\bar{N}(v)\frac{2I_1\eta_1\gamma_{ba}^2}{(\gamma_{ba}^2+(\Delta_1+k_1v)^2+2I_1\eta_1\gamma_{ba}^2)(\Delta_2+k_2v-i\gamma_{cb})}. \tag{8.3.4}$$

The imaginary part of $\tilde{\rho}_{cb}{}^{(2)}$ can be written as

$$Im\left(\tilde{\rho}_{cb}{}^{(2)}\right) = -\frac{\bar{N}(v)\Omega_2 I_1\eta_1\gamma_{ba}^2\gamma_{cb}}{(\gamma_{ba}^2+(\Delta_1+k_1v)^2+2I_1\eta_1\gamma_{ba}^2)[(\Delta_2+k_2v)^2+\gamma_{cb}{}^2]} \tag{8.3.5}$$

For small I_1, we can omit the power broadening term in the denominator and obtain

$$Im\left(\tilde{\rho}_{cb}{}^{(2)}\right) =$$

$$=-\bar{N}(v)\frac{I_1\eta_1\gamma_{ba}^2}{(\gamma_{ba}^2+(\Delta_1+k_1v)^2)}\cdot\frac{\Omega_2\gamma_{cb}}{\gamma_{cb}{}^2+(\Delta_2+k_2v)^2}. \tag{8.3.6}$$

Imaginary part of this matrix element gives rise to the absorption contribution to the $b \rightarrow c$ transition because of population changes caused by the $a \rightarrow b$ transition. It should be noted here that this term appears when the intermediate level population $\rho_{bb}{}^{(1)}$ is non-zero. Thus if because of collisions, laser field fluctuations or other incoherent processes, the term $\tilde{\rho}_{ba}{}^{(1)}$ goes to zero, we have an induced effect on the probe transition, if there is resonance in the pump transition. It is given by the product of two Lorentzians and can have maxima when the pump or probe transitions or both of them are on resonance.

The contribution arising from the coherent term $\tilde{\rho}_{ba}{}^{(1)}$ to dipolar matrix for the probe transition is

$$\left(\tilde{\rho}_{cb}{}^{(2)}\right)_{coh} = \frac{\bar{N}(v)\frac{\Omega_2}{2}(\frac{\Omega_1}{2})^2}{(\Delta_2+k_2v-i\gamma_{cb})(\Delta_1+\Delta_2+(k_1+k_2)v-i\gamma_{ca})}\,\frac{(\Delta_1+k_1v)+i\gamma_{ba}}{[(\Delta_1+k_1v)^2+\gamma_{ba}^2]}. \tag{8.3.7}$$

The imaginary part of $\left(\tilde{\rho}_{cb}{}^{(2)}\right)_{coh}$ has a complex dependence on the detuning terms. But in this term, it is observed that there can be maxima, when $\Delta_1 = 0$ or $\Delta_2 = 0$ or $\Delta_1 + \Delta_2 = 0$ or when all of them are zero for atoms of a particular velocity. In addition to the on-resonance pump or probe transitions there can be an additional effect when $\Delta_1 + \Delta_2 = 0$ or when $\omega_{ca} = \nu_1 + \nu_2$. Thus if the sum of the pump and probe frequencies is equal to the energy difference between the two levels, we can have a **two-quantum or two-photon transition**, even if the pump and probe transitions are not on resonance. It should be clear that the other case originating from the population transfer is also a two-photon process since it involves two quantum transitions although it needs the pump to be on resonance.

8.4 Doppler free two-photon spectroscopy

In the cascade process, we consider both the pump and probe transitions to be off-resonance, so that $\Delta_1 \neq 0$ $\Delta_2 \neq 0$, but $\Delta_1 + \Delta_2 = 0$. Thus the two-photon transition is near resonance. In this case, the two-step process has no contribution, because the large detuning of single photon process will not allow population buildup in the intermediate level.

If the detunings are large enough, so that

$$\Delta_1 \gg k_1 v \text{ and } \Delta_2 \gg k_2 v,$$

these detunings will dominate all other terms in the denominator of Eq. (8.3.7) and we can write

$$\left(\tilde{\rho}_{cb}{}^{(2)}\right)_{coh} =$$

$$= \frac{\bar{N}(v)\frac{\Omega_2}{2}(\frac{\Omega_1}{2})^2(\Delta_1 + \Delta_2 + (k_1 + k_2)v + i\gamma_{ca})/\Delta_1\Delta_2}{(\Delta_1 + \Delta_2 + (k_1 + k_2)v)^2 + {\gamma_{ca}}^2}. \quad (8.4.1)$$

The imaginary part of the above gives a Lorentzian line shape with width γ_{ca} and resonance at $\Delta_1 + \Delta_2 + (k_1 + k_2)v = 0$. It should be mentioned that we have assumed the field strength to be small so that terms with higher powers of $\mathcal{E}$ are neglected. We get resonance for the velocity satisfying

$$\Delta_1 + \Delta_2 = -(k_1 + k_2)v \quad (8.4.2)$$

For zero velocity group of atoms $\Delta_1 + \Delta_2 = 0$ or $\omega_{ca} = \nu_1 + \nu_2$. If $\nu_1 = \nu_2 = \nu$ we have a single radiation frequency and the resonance condition is

$$\omega_{ca} = 2\nu . \quad (8.4.3)$$

Hence for zero velocity group of atoms, we get a two photon resonance condition when the energy gap is twice the radiation frequency, if radiations have the same frequency. For all atoms having zero or non-zero velocity, if the laser radiations are moving in opposite directions (Fig. 8.2a) we have

$$k_1 = -k_2 = k \text{ or, } k_1 + k_2 = 0. \quad (8.4.4)$$

This leads to $\Delta_1 + \Delta_2 = 0$ for atoms of all velocities. Thus we have velocity independent resonance. All atoms within the Doppler contour will face the same excitation and so a very sharp Doppler free resonance is seen. This will exhibit a very narrow two-photon spectrum (Fig. 8.3). In this case the atoms get simultaneous Doppler shift to $\nu + kv$ and $\nu - kv$. They cancel each other. If the two

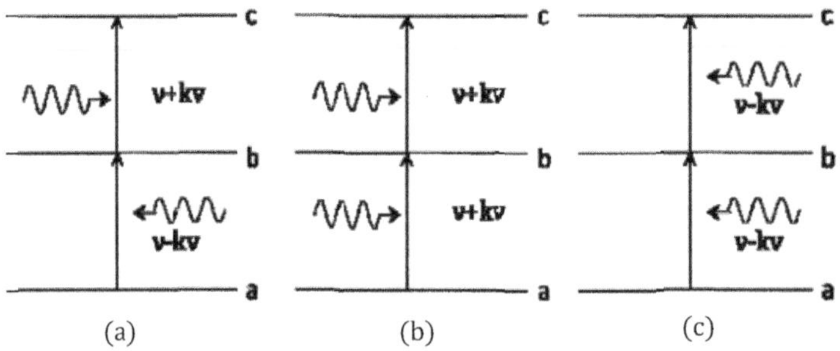

Fig. 8.2 Two waves of same frequency moving in (a)opposite directions, (b) both waves moving in forward directions and (c) both waves moving in the backward directions.

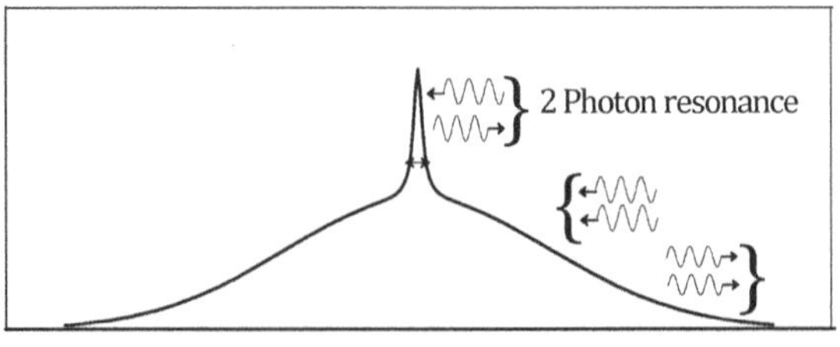

Fig.8.3 Doppler-free two photon resonance spectrum shown as a Lorentzian on a Gaussian pedestal arising from the two forward moving or the two backward moving waves.

photons moving in the same direction (Fig. 8.2b) are absorbed by the atom, the resonance condition is velocity dependent. For the wave moving in the forward direction and absorbing two photons we get

$$\omega_{ca} = 2(\, \nu + kv). \tag{8.4.5}$$

And for the wave moving in the backward direction (Fig. 8.2c) we get the resonance condition as

$$\omega_{ca} = 2\,(\nu - kv). \tag{8.4.6}$$

These resonances being velocity dependent will form the broad Doppler background as shown in Fig. 8.3 as a Gaussian profile. Thus we can get Doppler-free two photon spectroscopy in a three level atom with a single radiation moving in opposite directions. This may be analogous to saturation spectroscopy, but the fundamental difference is that the saturation spectroscopy occurs for two-level atom and appears as a hole on the Doppler background. In the present case of three level spectroscopy, a sharp peak appears on the Doppler profile. But it also has a Lorentzian line shape.

8.5 Dark state and Coherent Population Trapping

We have noticed that a three level system offers very exciting new possibilities that can be completely different from two-level atom spectroscopy. One remarkable feature is the formation of dark states in which two separate incident radiations can produce a quantum mechanical superposition state, where the atoms may get trapped and no transition is possible from that state. The phenomenon is called Coherent Population Trapping (CPT). This is very common for the $\Lambda -$ type systems of Fig. 8.1(c).

We denote the eigenstates as $|a\rangle$, $|b\rangle$ and $|c\rangle$. The unperturbed atomic Hamiltonian can be written as

$$H_{atom} = \hbar\omega_a|a\rangle\langle a| + \hbar\omega_b|b\rangle\langle b| + \hbar\omega_c|c\rangle\langle c|. \tag{8.5.1}$$

and the interaction Hamiltonian in the presence of the pump and probe field as described earlier is

$$H_{int} = -\frac{\hbar}{2}[\Omega_2 e^{-i\nu_2 t}|a\rangle\langle b| + C.C.] - \frac{\hbar}{2}[\Omega_1 e^{-i\nu_1 t}|b\rangle\langle c| + C.C.]. \tag{8.5.2}$$

We have assumed the rotating wave approximation and the electric fields are

$$\mathcal{E}_1 = \mathcal{E}_{10}\cos\nu_1 t, \tag{8.5.3}$$

$$\mathcal{E}_2 = \mathcal{E}_{20}\cos\nu_2 t. \tag{8.5.4}$$

The Rabi frequencies are the same as those defined earlier, we have excluded the z-dependence of the electric field. We also assumed that the transition $c \leftrightarrow a$ is not dipole allowed.

The atomic wave function can be written as

$$|\psi\rangle = c_a|a\rangle + c_b|b\rangle + c_c|c\rangle. \tag{8.5.5}$$

From the Schrödinger equation

$$i\hbar|\dot{\psi}\rangle = (H_{atom} + H_{int})\,|\psi\rangle$$

we can get the rate equations for the c-coefficients as

$$\dot{c}_a = -i\omega_a c_a + \frac{i\Omega_2}{2} e^{i\nu_2 t} c_b, \tag{8.5.6}$$

$$\dot{c}_b = -i\omega_b c_b + \frac{i\Omega_2}{2} e^{-i\nu_2 t} c_a + \frac{i\Omega_1}{2} e^{-i\nu_1 t} c_c, \tag{8.5.7}$$

$$\dot{c}_c = -i\omega_c c_c + \frac{i\Omega_1}{2} e^{i\nu_1 t} c_b. \tag{8.5.8}$$

We define

$$c_a = C_a e^{-i\omega_a t}, \tag{8.5.9}$$

$$c_b = C_b e^{-i\omega_b t}, \tag{8.5.10}$$

$$c_c = C_c e^{-i\omega_c t}. \tag{8.5.11}$$

The rate equations for the C-coefficients are

$$\dot{C}_a = +\frac{i\Omega_2}{2} C_b, \tag{8.5.12}$$

$$\dot{C}_b = \frac{i\Omega_2}{2} C_a + \frac{i\Omega_1}{2} C_c, \tag{8.5.13}$$

$$\dot{C}_c = \frac{i\Omega_1}{2} C_b. \tag{8.5.14}$$

An adiabatic solution will lead to zero probability for the atoms in the upper state so that we get

$$C_b = 0$$

and

$$\frac{C_a}{C_c} = -\frac{\Omega_1}{\Omega_2}.$$

Using the normalization condition,

$$|C_a|^2 + |C_b|^2 + |C_c|^2 = 1$$

we get,

$$C_a = \frac{\Omega_1}{\sqrt{|\Omega_1|^2 + |\Omega_2|^2}},$$

$$C_c = -\frac{\Omega_2}{\sqrt{|\Omega_1|^2 + |\Omega_2|^2}}.$$

We have omitted an overall phase term. The time dependent wave function can be written as

$$|\psi(t)\rangle = \frac{\Omega_1 e^{-i\omega_a t}|a\rangle - \Omega_2 e^{-i\omega_c t}|c\rangle}{\sqrt{|\Omega_1|^2 + |\Omega_2|^2}}. \tag{8.5.15}$$

The state described by the wave function in Eq. (8.5.15) is called a dark state. This state does not contain the upper state $|b\rangle$. The atom remains confined or trapped in the lower state. This arises because of the destructive quantum interference between the two transition pathways $a \rightarrow b$ and $c \rightarrow b$. The two terms in the wave function lead to the probability amplitude of the two transitions $a \rightarrow b$ and $c \rightarrow b$. In the case of destructive interference, the combined probability for the transitions from the lower states $|a\rangle$ and $|c\rangle$ is zero. So the particle remains trapped in the lower states. The two states together form the dark state and the probability of the two individual states in the dark state depends on their respective Rabi frequencies. The state is also called trapped state or non-coupled state, as it remains uncoupled with the upper state.

We can also understand the dark state as superposition of states into dark and bright states. For the Λ-type three level system of Fig 8.1(c), we can assume the field strengths to be equal. In that case the Rabi frequencies are the same so that $\Omega_1 = \Omega_2 = \Omega$. We can consider the superposition states instead of the original states as the lower states get coupled by the simultaneous interaction of the two fields. Hence the new basis states can be

$$|\psi\rangle_{bright} = \frac{1}{\sqrt{2}}[|a\rangle + |c\rangle]. \quad (8.5.16)$$

$$|\psi\rangle_{dark} = \frac{1}{\sqrt{2}}[|a\rangle - |c\rangle] \quad (8.5.17)$$

and

$$|\psi\rangle_{\text{upper}} = |b\rangle$$

It is found that the atomic state prepared in the **Bright State** (also called coupled state) can interact with the radiation and absorb light. Whereas the state prepared in the **Dark State** (also called non-coupled state) cannot couple to the radiation as the transition probability from this state to the upper state is zero. Hence the medium will be transparent for the wave corresponding to the transition $|\psi\rangle_{dark}$ to the $|\psi\rangle_{\text{upper}}$ state. The dark states are important in Velocity Selective Coherent Population Trapping (VSCPT)

The coupled state can absorb photons and transfer them to the non coupled state through spontaneous emission, so all atoms could be trapped into the $|\psi\rangle_{dark}$ state. This phenomenon is called **Coherent Population Trapping (CPT).**

When $\Omega_1 \approx \Omega_2$, we have complete cancellation of absorption leading to **Coherent Population Trapping (CPT).** It may be noted from Eq. (8.5.14) that for complete cancellation we need $\langle b|\mu|\psi_{dark}\rangle = 0$. This is possible for same laser intensity of the two radiations (see Problem 8.1).

If $\Omega_1 << \Omega_2$; reduction of absorption is possible leading to **Electromagnetically induced transparency (EIT).** We will show that in the case of EIT, the atomic state evolves into the dark state that is transparent to light.

The basic difference between CPT and EIT is that in case of CPT, we start with an atom that can be in dark and bright states. This can result by the combined action of applied electromagnetic field and optical pumping. In the case of EIT the atom evolves into a dark state by the simultaneous action of the strong pump and weak probe field and is not initially prepared in that state.

8.6 Electromagnetically Induced Transparency

We consider a closed three level system described in Fig. 8.4. The levels a and b are coupled by a weak probe field of frequency ν_1. The levels b and c are coupled by a strong pump (also called coupling or driving) field of frequency ν_2. Both of them can have small detuning.

At Raman resonance i.e. $\hbar(\nu_1 - \nu_2) = E_a - E_c$, a coherent superposition of the ground states creates a dark and a bright state. The dark state leads to transparency. It results from destructive interference between the two pathways to the excited state.

The Hamiltonian of this Λ −type atom interacting with the two fields is the same as given in Eq. (8.5.2-4). The wave function of the atom is described in

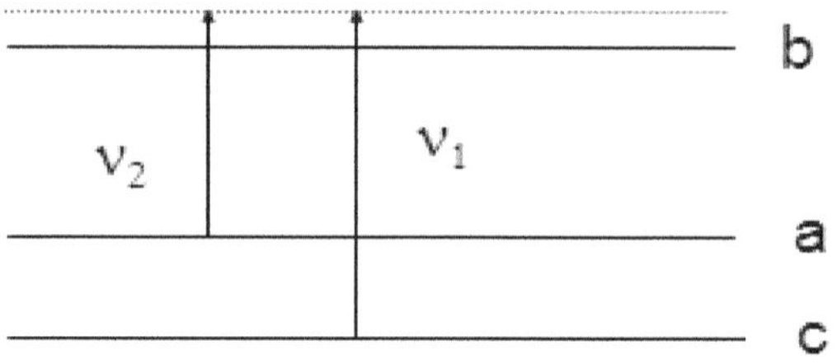

Fig. 8.4 Three level atom interacting with two laser radiations. If they are both detuned the upper level for the transition is shown as a fictitious level (dotted line) that is very close to the upper level b. The transitions satisfy the Raman condition.

terms of the c-coefficients and the Schroedinger equation leads to the rate equations as given by the Eq. (8.5.7). Introducing the density matrix elements $\rho_{nm} = c_n c_m^*$ we can obtain the density matrix equations for the off-diagonal elements following the method described in Chapter 3.

$$\dot{\rho}_{bc} = -i\omega_{bc}\rho_{bc} + i\frac{\Omega_1}{2}e^{-i\nu_1 t}(\rho_{cc} - \rho_{bb}) + i\frac{\Omega_2}{2}e^{-i\nu_2 t}\rho_{ac}, \quad (8.6.1)$$

$$\dot{\rho}_{ba} = -i\omega_{ba}\rho_{ba} + i\frac{\Omega_2}{2}e^{-i\nu_2 t}(\rho_{aa} - \rho_{bb}) + i\frac{\Omega_1}{2}e^{-i\nu_1 t}\rho_{ca}, \quad (8.6.2)$$

$$\dot{\rho}_{ac} = -i\omega_{ac}\rho_{ac} - i\frac{\Omega_1}{2}e^{-i\nu_1 t}\rho_{ab} + i\frac{\Omega_2}{2}e^{i\nu_2 t}\rho_{cb}, \quad (8.6.3)$$

where $\omega_{nm} = \omega_n - \omega_m, (n, m = a, b, c)$ are the resonance frequencies and Ω_1 and Ω_2 are the Rabi frequencies of the probe and pump transitions respectively. We have written equations for the off-diagonal elements only since these equations are necessary to obtain absorption and dispersion line profiles for three level atom in the presence of a strong pump transition. We have not included the decay terms. Introducing exponential decay γ_1, γ_2 and γ_3 for the elements ρ_{bc}, ρ_{ba} and ρ_{ac} respectively we can write the rate equations as

$$\dot{\rho}_{bc} = -(i\omega_{bc} + \gamma_1)\rho_{bc} + i\frac{\Omega_1}{2}e^{-i\nu_1 t}(\rho_{cc} - \rho_{bb}) + i\frac{\Omega_2}{2}e^{-i\nu_2 t}\rho_{ac}, \quad (8.6.4)$$

$$\dot{\rho}_{ba} = -i(\omega_{ba} + \gamma_2)\rho_{ba} + i\frac{\Omega_2}{2}e^{-i\nu_2 t}(\rho_{aa} - \rho_{bb}) + i\frac{\Omega_1}{2}e^{-i\nu_1 t}\rho_{ca}, \quad (8.6.5)$$

$$\dot{\rho}_{ac} = -i(\omega_{ac} + \gamma_3)\rho_{ac} - i\frac{\Omega_1}{2}e^{-i\nu_1 t}\rho_{ab} + i\frac{\Omega_2}{2}e^{i\nu_2 t}\rho_{cb}. \quad (8.6.6)$$

As mentioned earlier the probe field is much weaker than the pump field i.e. $|\Omega_1| \ll |\Omega_2|$, for calculation of atomic coherence ρ_{bc} we shall consider the probe Rabi frequency Ω_1 to first order only and retain terms to all orders of Ω_2.

We assume the initial condition that all the atoms are in the lowest state, so

$$\rho_{cc}(0) = 1, \; \rho_{aa}(0) = 0, \quad \rho_{bb}(0) = 0 \quad (8.6.7)$$

Under the condition $|\Omega_1| \ll |\Omega_2|$, Eq (8.5.) shows that the dark state approaches the state $|c\rangle$. In this case all the atomic population will be in the state $|c\rangle$ in the first order of approximation as mentioned above, hence

$$\rho_{cc} \cong 1, \; \rho_{aa} \cong 0, \; \rho_{bb} \cong 0. \quad (8.6.8)$$

Because of the weak probe field we also have $\rho_{ab} \cong 0$, in the first order. But ρ_{bc} and ρ_{ca} will be non-zero in the first order.

Substituting these approximate values in the density matrix equations we get for ρ_{bc} and ρ_{ca} the following rate equations

$$\dot{\rho}_{bc} = -(i\omega_{bc} + \gamma_1)\rho_{bc} + i\frac{\Omega_1}{2}e^{-i\nu_1 t} + i\frac{\Omega_2}{2}e^{-i\nu_2 t}\rho_{ac}, \tag{8.6.9}$$

$$\dot{\rho}_{ac} = -i(\omega_{ac} + \gamma_3)\rho_{ac} + i\frac{\Omega_2}{2}e^{i\nu_2 t}\rho_{bc}. \tag{8.6.10}$$

Substituting the density matrix elements as

$$\tilde{\rho}_{bc} = e^{i\nu_2 t}\rho_{bc}\,, \tag{8.6.11}$$

$$\tilde{\rho}_{ac} = e^{i(\nu_1-\nu_2)t}\rho_{ac} \tag{8.6.12}$$

we can get the rate equations as

$$\dot{\tilde{\rho}}_{bc} = i(\delta_1 + i\gamma_1)\tilde{\rho}_{bc} + i\frac{\Omega_1}{2} + i\frac{\Omega_2}{2}\tilde{\rho}_{ac}, \tag{8.6.13}$$

$$\dot{\tilde{\rho}}_{ac} = i[(\delta_1 - \delta_2) + i\gamma_3)]\tilde{\rho}_{ac} + i\frac{\Omega_2}{2}\tilde{\rho}_{bc}. \tag{8.6.14}$$

We have defined the detunings as

$$\delta_1 = \nu_1 - \omega_{bc} \tag{8.6.15}$$

$$\delta_2 = \nu_2 - \omega_{ba} \tag{8.6.16}$$

We consider slowly varying approximation under the adiabatic condition so that we can set the time derivatives to zero. Hence

$$(\delta_1 + i\gamma_1)\tilde{\rho}_{bc} + \frac{\Omega_2}{2}\tilde{\rho}_{ac} = -\frac{\Omega_1}{2}, \tag{8.6.17}$$

$$[(\delta_1 - \delta_2) + i\gamma_3)]\tilde{\rho}_{ac} + \frac{\Omega_2}{2}\tilde{\rho}_{bc} = 0. \tag{8.6.18}$$

Solving these equations we can get the coherence for the probe transition to first order

$$\tilde{\rho}_{bc} = \frac{(\frac{\Omega_1}{2})[(\delta_1 - \delta_2) + i\gamma_3)]}{(\frac{\Omega_2}{2})^2 - (\delta_1 + i\gamma_1)[(\delta_1 - \delta_2) + i\gamma_3)]}. \tag{8.6.19}$$

If the pump frequency is exactly tuned to resonance $\delta_2 = 0$.

As mentioned in Chapter 3 (Eq. 3.5.24), the total polarization associated with the probe transition can be written as

$$P(t) = N_0\mu_{bc}(\rho_{bc} + \rho_{cb}) = N_0\mu_{bc}(\tilde{\rho}_{bc}e^{i\nu_2 t} + \tilde{\rho}_{cb}e^{-i\nu_2 t}). \tag{8.6.20}$$

N_0 is the number of atoms per unit volume and μ is the transition matrix element for the probe transition. In terms of the complex susceptibility we can also write the time dependent polarization as

$$P(t) = \frac{1}{2}\epsilon_0\mathcal{E}_0\,[\chi e^{i\nu_2 t} + \chi^* e^{-i\nu_2 t}]. \tag{8.6.21}$$

Hence

$$\chi = 2\frac{N_0\mu_{bc}}{\epsilon_0 \mathcal{E}_0}\tilde{\rho}_{bc} = 2\frac{N_0\mu_{bc}^2}{\epsilon_0\hbar\Omega_1}\tilde{\rho}_{bc}.$$

Substituting $\tilde{\rho}_{bc}$ we get

$$\chi = \frac{N_0\mu_{bc}^2}{\epsilon_0\hbar}\frac{[\delta_1 + i\gamma_3]}{(\frac{\Omega_2}{2})^2 - (\delta_1 + i\gamma_1)[(\delta_1 + i\gamma_3)]}. \tag{8.6.22}$$

Hence the real and imaginary parts of the susceptibility are

$$\chi' = \frac{N_0\mu_{bc}^2}{\epsilon_0\hbar}\frac{\delta_1[(\frac{\Omega_2}{2})^2 - \delta_1^2 - \gamma_3^2]}{[(\frac{\Omega_2}{2})^2 - \delta_1^2 + \gamma_1\gamma_3]^2 + \delta_1^2(\gamma_1 + \gamma_3)^2}, \tag{8.6.23}$$

$$\chi'' = \frac{N_0\mu_{bc}^2}{\epsilon_0\hbar}\frac{\gamma_3[(\frac{\Omega_2}{2})^2 + \gamma_1\gamma_3] + \delta_1^2\gamma_1}{[(\frac{\Omega_2}{2})^2 - \delta_1^2 + \gamma_1\gamma_3]^2 + \delta_1^2(\gamma_1 + \gamma_3)^2}. \tag{8.6.24}$$

Fig. 8.5 shows the plot of χ' and χ'' against detuning in MHz units. Both the dispersion and absorption parts are zero at zero detuning. Thus the medium is transparent at resonance of probe frequency. This transparency is induced by the presence of the coupling light source. The phenomenon is called Electromagnetically Induced Transparency (EIT).

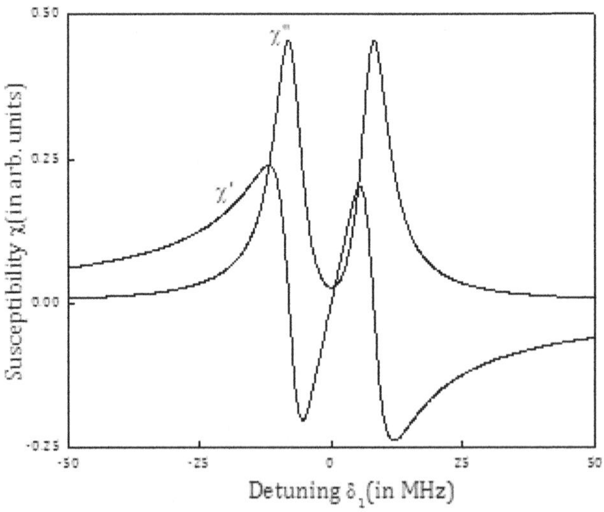

Fig. 8.5 Real and imaginary parts of susceptibility plotted against detuning.

To explain the origin of EIT we analyze the result at zero detuning of the probe. We must note here that in this section we have considered the coupling or pump frequency to be on resonance. When $\delta_1 = 0$, we get

$$\chi'(\delta_1 = 0) = 0, \tag{8.6.25}$$

$$\chi''(\delta_1 = 0) = \frac{N_0 {\mu_{bc}}^2}{\epsilon_0 \hbar} \frac{\gamma_3}{[(\frac{\Omega_2}{2})^2 + \gamma_1\gamma_3]}. \quad (8.6.26)$$

It may be noted that γ_3 is the dephasing rate of the dipole forbidden transition, so it is very small. Hence the absorption coefficient can be very small even if it is not exactly zero.

In the strong coupling condition

$$(\frac{\Omega_2}{2})^2 \gg \gamma_1\gamma_3.$$

Hence

$$\chi''(\delta_1 = 0) \cong \frac{N_0 {\mu_{bc}}^2}{\epsilon_0 \hbar} \frac{4\gamma_3}{{\Omega_2}^2}. \quad (8.6.27)$$

It goes to zero when γ_3 is small and Ω_2 is large. Thus a strong coupling field can restrain absorption at probe resonance and the medium becomes transparent. The physical process involved is that the atom goes to the dark state and absorption is not possible.

8.7 EIT and Optical Pumping in multilevel atoms

In Chapter 4, we discussed the atomic energy levels of Rubidium atom. We showed how the hyperfine energy levels modify the saturation absorption spectra of Rb atom. In this section, we show how an experimental spectrum on EIT of Rb gets modified because of the closely spaced energy levels. We have, so far, discussed the fundamental process involved in the EIT of three level systems. But the alkali atoms have closely spaced energy levels that do not allow observation of spectra in pure three level systems in usual experiments.

We must note the following points in the case of observation of pump probe spectra in alkali atoms.

The experiments are usually performed in gas phase at room temperature.
The observed spectrum is, therefore, a convolution of a Doppler broadened Gaussian and the Lorentzian line shapes, as we discussed in the former section. The Gaussian line profile may be subtracted from the observed curve to recover the Lorentzian shape.
The Doppler broadening at normal temperature is usually of the order of a few hundred MHz. So it is much larger than the hyperfine splitting. Hence a few hyperfine transitions are within the envelope of one Doppler profile.
As discussed in Chapter 3, there can be optical pumping of atoms from one lower level to the other. Such optical pumping can be velocity selective since, in the course of scanning the probe frequency with the fixed pump frequency, the probe radiation will have resonance with other nearby hyperfine levels. This is called Velocity Selective Optical Pumping (VSOP).
EIT is observed at the centre of one such VSOP when the Raman condition is satisfied.

The spectra of ^{87}Rb will be illustrated in the following. As mentioned earlier for the D_2 transition ($^2S_{1/2}$ –$^2P_{3/2}$), we have to consider a six-level atom (Fig. 8.6). The lower level for the pump or control transition is $|b\rangle$ (F =2) and the same for the probe transition is $|a\rangle$ (F =1). The upper state can be any one of the states that is allowed by the selection rule, $\Delta F = 0$. We consider the control frequency ν_c to be fixed and it is detuned at a frequency between the F' = 2 and 1 levels (shown as a dotted level). The probe laser frequency ν_p is tuned in the experiment and it can have resonances for the transitions F=1 to F' = 0,1,2.

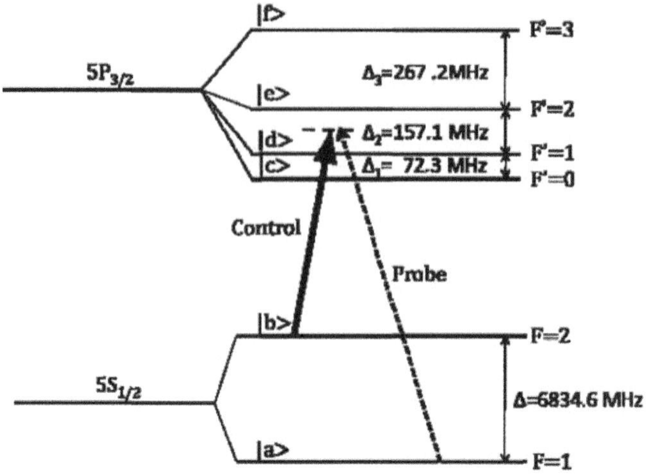

Fig. 8.6 Five level Λ configuration of pump probe spectrum of [87]Rb-D2 transitions, The gound state hyperfine splitting is denoted as Δ and the upper state splitting are Δ_i $(i = 1,2,3)$.

The probe spectra are studied in the presence of a fixed control frequency. Hence we effectively consider a five-level atom. In the gas phase spectroscopy all the three probe transitions will be enveloped by the Doppler profile that has a full width at half maximum of 450 MHz. The pump transition, although detuned from any transition, will have resonances with all the allowed transitions to the levels F' = 1,2,3 because the atom sees a Doppler shifted frequency. Hence pump transitions can occur for groups of atoms selected by their velocities that are non-zero. During the process of tuning the probe laser will sweep across the upper levels F=0,1,2 that may be on resonance with the selected group of atoms having appropriate velocities so that they are on resonance with the control laser.

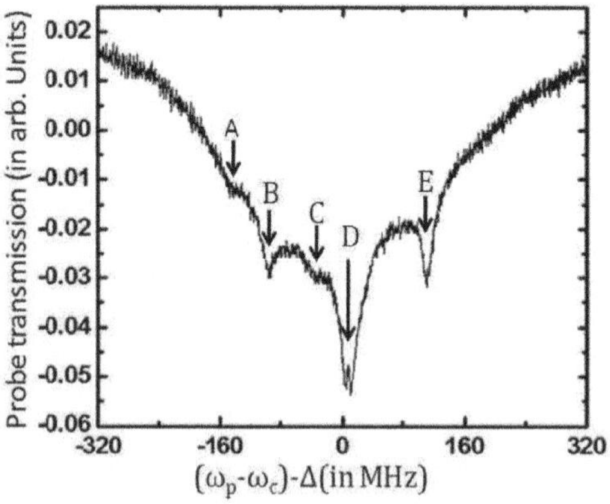

Fig. 8.7 Observed velocity selective optical pumping and EIT observed in D_2 transition of [87]Rb atom. The peak (D) at the centre of the VSOP is EIT.

[Reprinted with the permission from Journal of Physics B (Ref. 8)

Since the pump laser will be on resonance with the transitions to the states F' =1, 2, 3 for non-zero velocity atoms, a large number of atoms will be pumped to these levels. Eventually they will decay by spontaneous emission to the levels F=1, 2. Since the control laser is much stronger than the probe laser there will be large accumulation of atoms in the ground level F=1. This is velocity selective optical pumping (VSOP). This increases the population difference for the probe transition thereby enhancing the absorption intensity at these resonances. Such enhanced absorptions are shown as A, B, C, D and E in the spectra of Fig. 8.7.

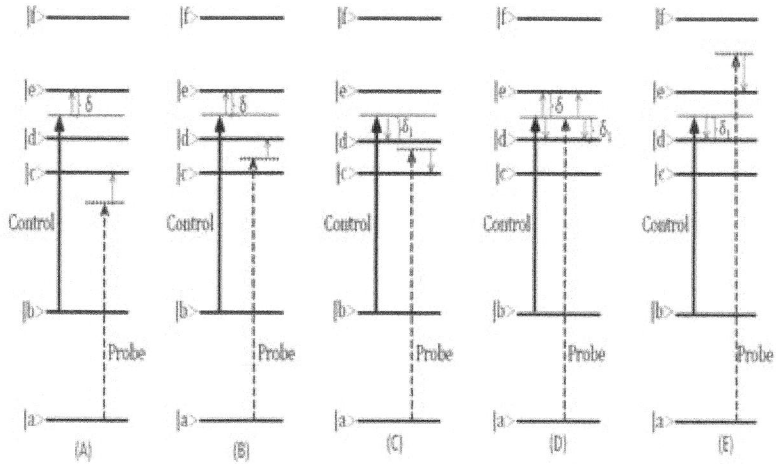

Fig. 8.8 Energy level diagram for six-level Rb atom depicts the Velocity Selective Control and Probe transitions leading to VSOP and EIT.

[Reprinted with the permission from Journal of Physics B (Ref. 8)

It is interesting to note that the separation of these peaks can be explained on the basis of hyperfine splitting. We denote the atomic resonance frequencies as $\omega_{ij}, i = a, b;\ j = c, d, e$ (Fig. 8.8). The radiation wave vectors are $\vec{k}_c$ and $\vec{k}_p$ for the control and probe lasers respectively. The control laser is locked at a frequency that is red detuned with respect to the $b \to e$ transition. Hence $\nu_c = \omega_{be} - \delta$, where $\delta = \vec{k}_c \cdot \vec{v}_1$ and $\vec{v}_1$ is the velocity of the atoms that absorb the pump radiation. Since in this case, the laser radiations have almost the same frequency and they are moving in parallel directions we can assume $\vec{k}_c \approx \vec{k}_p$. For the same velocity group of atoms, the probe radiation will be on resonance with the transitions $a \to c$, $a \to d$ and $a \to e$ at the respective frequencies $\nu_p = \omega_{ac} - \delta$, $\nu_p = \omega_{ad} - \delta$ and $\nu_p = \omega_{ae} - \delta$. When the probe laser frequency attains the value $\omega_{ae} - \delta$, both the pump and probe lasers have the same upper level at resonance for this velocity group. In this case the transitions undergo a destructive interference as discussed in the previous section. Hence we can see the medium to be more transparent. At this transition we also have a VSOP and the Electromagnetically Induced Transparency (EIT) is sitting on the VSOP peaks. For the other two transitions, the Raman conditions are not satisfied and we cannot expect to see EIT, but we have the VSOP peaks at A and B.

Similarly the control laser frequency is blue detuned from the transition $b \to d$, hence $\nu_c = \omega_{bd} + \delta_1$, where $\delta + \delta_1 = \Delta_2$. Thus the control laser is tuned to resonance for atoms of the velocity group satisfying $\delta_1 = \vec{k}_c \cdot \vec{v}_2$. For the same velocity group of atoms the probe beam will be on resonance with the transitions $a \to c$, $a \to d$ and $a \to e$ and the respective frequencies are $\nu_p = \omega_{ac} + \delta_1$, $\nu_p = \omega_{ad} + \delta_1$ and $\nu_p = \omega_{ae} + \delta_1$. When the upper level for the probe transition is e, both the control and probe transitions will have the same upper level and Raman condition for EIT is fulfilled (D in Fig. 8.8). For the other two resonances we cannot have EIT and we get VSOP peaks at C and E (Fig. 8.8). Hence the separation of the observed VSOP peaks depends on the hyperfine energy differences of the ground and excited states (Problem 8.4). Hence all the five VSOP signals and the EIT signal observed in Fig. 8.7 can be explained.

Problem

8.1 From Eq. (8.5.14) for the wave function of the dark state (when $\Omega_1 \approx \Omega_2$) show that the transition probability to the upper state can be zero if the laser intensities are the same.

8.2 Show that for the dark state given by Eq. (8.5.11) the transition probability to the upper state is not completely zero even if $\Omega_1 \approx \Omega_2$. Give physical interpretation of the result.

8.3 Obtain an expression of susceptibility when
$\delta_1 = \delta_2 = \delta \neq 0$
Give so that both the radiations are not on resonance. Find the real and imaginary parts.

8.4 Show that for the peaks A, B, C, D, E in Fig. 8.6 the difference between the fixed control frequency and the probe frequencies are as follows
$\nu_p - \nu_c = \Delta - \Delta_1 - \Delta_2$
$\nu_p - \nu_c = \Delta - \Delta_2$
$\nu_p - \nu_c = \Delta - \Delta_1$
$\nu_p - \nu_c = \Delta$
$\nu_p - \nu_c = \Delta + \Delta_2$
Where Δ, Δ_1, Δ_2 are the hyperfine splitting as shown in Fig. 8.5.

Further reading

1. P. Meystre and M. Sargent III, Elements of Quantum Optics, Springer, Berlin, 2007.
2. M. O. Scully and M.S. Zubairy, Quantum Optics, Cambridge University Press, Cambridge, 1997.
3. S. C. Rand, Non-linear and Quantum Optics, Oxford, New York, 2010.
4. S. Harris, Physics Today, 1997, vol. 50, pp.36.
5. K. Hakuta, L. Marmet and B. P. Stoicheff, Phys. Rev. Lett., 1991, vol 66, pp. 596.
6. K.-J. Boller, A. Imamoglu and S. E. Harris, Phys. Rev. Lett., 1991, vol 66, pp. 2593.
7. E. Arimondo, Progress in Optics, 1996, Elsevier, Amstardam.
8. S. Chakraborty, A. Pradhan, B. Ray and P. N. Ghosh, J. Phys. B, 2005, vol. 38, pp. 4321-27.

Chapter 9
Laser Cooling and Bose Einstein Condensation

9.1 Basic phenomena of atom cooling and trapping

9.1.1 Temperature

When an object is hot, the atoms inside it are moving fast in random directions and when it gets cold, they are moving slowly. Our body interprets that random atomic motion into our feeling of hot and cold, and a thermometer presents that atomic motion in terms of a certain number of degrees. If the object is a solid the atoms are oscillating back and forth, and if it is a gas like the air, the atoms are moving around.

In a group of atoms there is always a whole range of speeds, but while the speed of one atom changes, the average of all of them may not change. Each time an atom slows down; there may be another that speeds up. So temperature is really describing the range of speeds of the bunch of atoms together.

The coldest place in nature is in the depths of interstellar space. There, it is nearly 3 degrees Kelvin. It was shown that the heat left over from the Big Bang is everywhere, and it prevents the temperature in space from going down to lower than 3 degrees Kelvin. Measurement of this temperature confirms that the Big Bang actually happened for creation of the Universe. The range of temperature that can actually be achieved in the laboratory covers wide range of nearly 10^{12} K. Fig. 9.1 gives an idea of temperature range in logarithmic scale that can be accessed. The highest temperature that has been recorded is close to the temperature on the surface of the Sun. This has been attained in the laboratory by the method of laser induced internal friction.

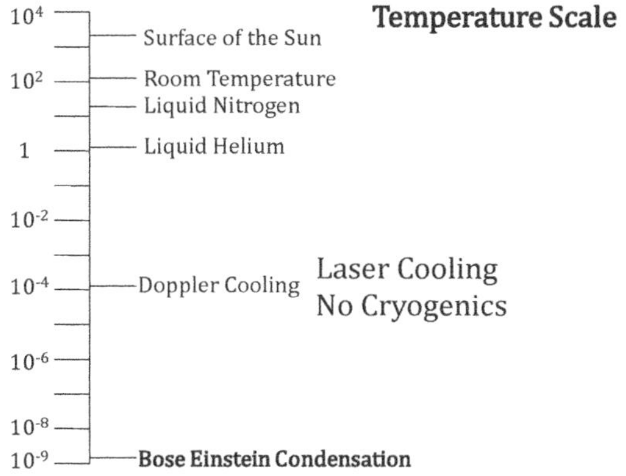

Fig. 9.1 Temperature range in Kelvin in logarithmic scale.

However, we can do much better than nature for getting things colder. Over the last hundred years, we have been able to build refrigerators that cool to lower than 3 degrees above Absolute Zero. More recently, we have been able to even get lower than 1 mK. However, it was a big step when Cornell and Wieman in Boulder and Ketterle in Boston cooled a small sample of atoms down to 100 nK. That was necessary in order to observe Bose-Einstein condensation.

9.1.2 Kinetics of atomic motion

It is known from kinetic theory of gases that at normal temperature and pressure the atoms execute random motions and the root mean square velocity is given by

$$v_{rms} = \sqrt{\frac{3k_BT}{M}} \qquad (9.1.1)$$

where T is the absolute temperature, M is the atomic mass. If the gas phase atoms are confined in a container, the atomic motion can be described by the arrows as shown in Fig. 9.2. In statistical mechanics, temperature is defined as a macroscopic quantity, which expresses an average velocity of atoms in terms of v_{rms}. No atom in the gas phase can have zero velocity or can be at rest. This is also not allowed in quantum mechanics. The root mean square velocity cannot be zero since absolute zero of temperature cannot be attained according to the laws of thermodynamics. The gas phase atoms can be cooled following the gas laws. In order to cool the gas we can cool the container. In that case the atoms will move towards the surface of the container and will condense on the walls. This is not cooling of the gas phase atoms. In this case, the atoms will freeze on the walls of the container. By cooling of gas phase atoms we mean to slow down the atoms thereby reducing their kinetic energies. The rms velocity will then be reduced and hence the temperature is lower.

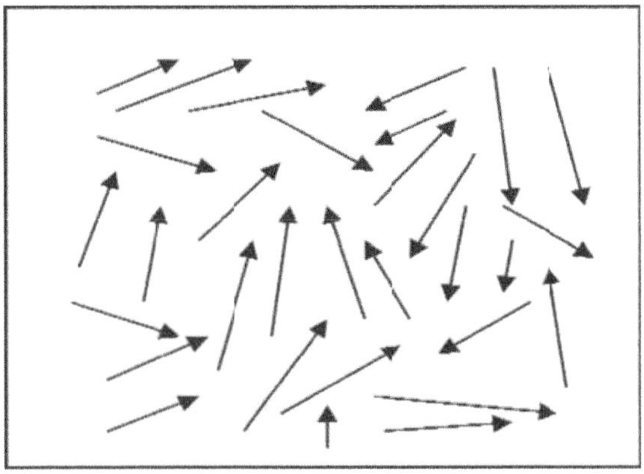

Fig. 9.2 Atomic motion at normal temperature is shown by arrow; the length of the arrow indicates the magnitude of the velocities. The directions are shown by the orientation of the arrows

9.1.3 Cooling and trapping of gas phase atoms

The aim of atom cooling experiments is to have a condition of the container gases as shown in Fig. 9.3. This is a case of dilute gas of atoms which have been decelerated and are still moving at random as a gaseous system. Hence rms velocity and consequently the temperature will be lower. It is necessary that the gas should be very dilute and that needs ultra-high vacuum. If the pressure is high, the atomic density will be large and inter-atomic forces may lead to coupling and bond formation at very low temperature. These bonded atoms may then serve as nuclei for condensation of the atoms. In the case of atom cooling, we need to decelerate the atoms to lower velocities. This has to be done for each atom individually. In a dilute and cooled gas it is necessary to confine the atoms in a small region of space so that experiments can be done on a group of such cooled atoms. This process is

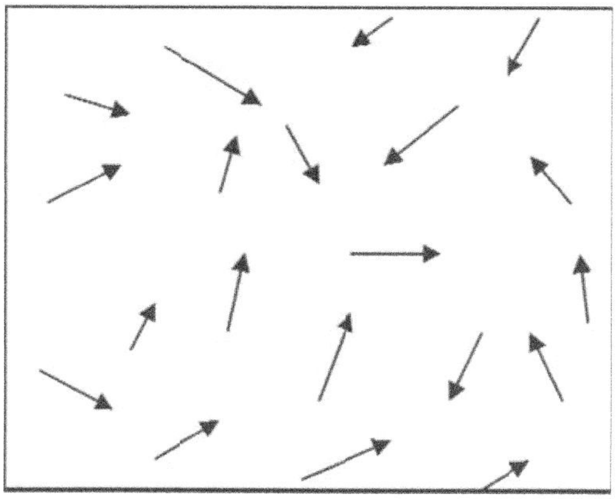

Fig. 9.3 Atomic velocities are low as indicated by the length of the vectors. The gas is also dilute so that the number density is low.

called trapping. We show the trapped atoms in Fig. 9.4. Normally the velocity of atoms that are of the order of 800-900 m/sec at room temperature are decelerated to 2 m/sec at a low temperature of nearly 100 micro-Kelvin (μ K). The area of confinement has a diameter of the order of 0.1 mm.

We shall discuss effective application of low power laser radiations for cooling and use of magnetic field gradient for trapping the atoms. Thus we discuss the basic principle of operation of Magneto-Optic Traps (MOT). For cooling of atomic motion, we need **velocity dependent damping force.**

For the purpose of trapping, we need **position dependent restoring force.**

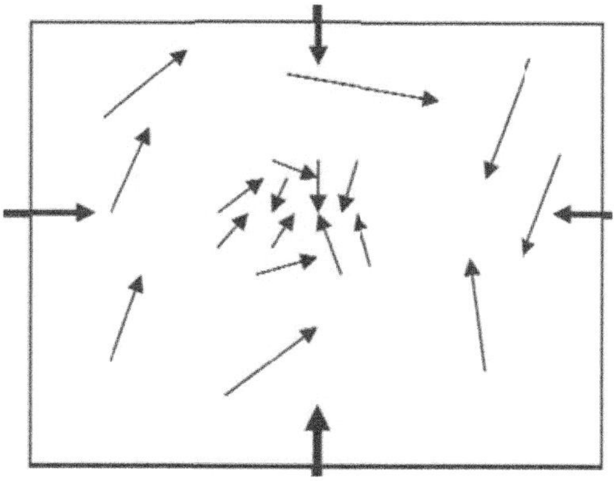

Fig. 9.4 Cooled low velocity atoms are pushed to the centre of the container thereby trapping them

9.2 Doppler Cooling

Schawlow first showed in the seventies that the optical force of laser radiation can be used to decelerate an atom. This can be accomplished by the absorption of a photon by an atom through a process of stimulated absorption if the laser radiation has a frequency resonant with the energy level difference of the atom. As the photon is absorbed, the momentum of the photon disappears since the photon does not exist. This can be considered as the result of a collision of a single atom with a single photon. In this case the law of conservation of linear momentum should hold. If an atom absorbs a photon coming from the opposite direction, according to the law of conservation of momentum, the momentum of the atom will be reduced since the atom receives a negative momentum from

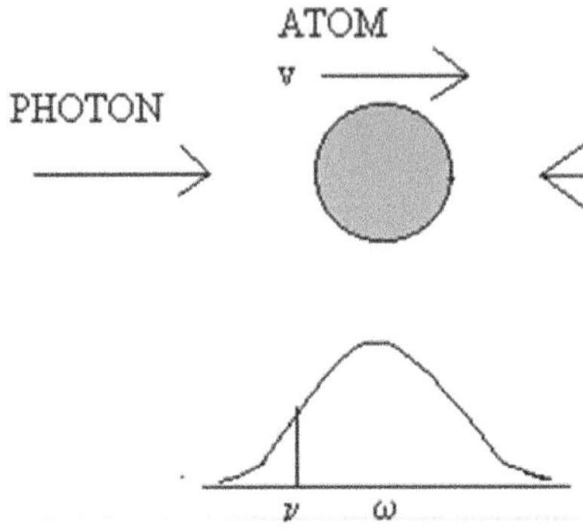

Fig.9.5 Basic process of Doppler cooling

the incident photon as the atom and the photon are moving in opposite directions (Fig. 9.5). Thus the atom velocity may be reduced resulting in a deceleration of the atom. If the atom absorbs a photon moving in the same direction as the atom, it may be accelerated on the basis of the same argument. The laser beam moving from two opposite directions along a particular path can be absorbed by two different atoms that are moving in opposition to each other. We show below that in the process of Doppler cooling, we can create a situation in which the atom absorbs the photon coming from the opposite direction. Thus each atom may be decelerated in the process. Considering three mutually perpendicular directions atomic velocity components in all three directions may be reduced.

In this process, three dimensional Doppler cooling can effectively reduce all the velocity components of the atoms in the three directions. However, there is a limit to this process of cooling. We have considered only the conservation of momentum. But each process of absorption or emission of a photon by the atom is also associated with a change of energy. In order to understand this process we need a theoretical discussion on the forces acting on the atom.

The process of laser cooling of atoms can be understood with the help of the momentum change. If a radiation with frequency ν and wave vector $\vec{k}$ falls on an atom with discrete energy levels E_1 and E_2 such that $\nu < \frac{E_2-E_1}{\hbar} = \omega$, the photon is absorbed by the atoms with velocity $\vec{v}$ such that they satisfy the condition that the Doppler shifted photon frequency is equal to the transition frequency or $\nu - \omega = \hbar\,\vec{k} \cdot \vec{v}$ and detuning or the frequency difference is within the Doppler broadening region or the radiation has a small red detuning. When an atom absorbs a photon it will gain a momentum $\hbar\vec{k}$. If an atom of mass m moving with a velocity $\vec{v}$ absorbs a photon coming from the opposite direction (Fig. 9.5), the law of conservation of linear momentum dictates that the total momentum of the atom-photon system should remain unchanged. Since the two momenta are in opposite directions we can write $m\vec{v} - \hbar\vec{k} = m\vec{v'}$. Hence the atom will be decelerated since $v' < v$.

In the case of Doppler cooling the laser frequency is set to a value lower than the resonance frequency. When the photon is moving in the direction opposite to the laser frequency the atom will see a higher frequency at $\nu(1 + \frac{v}{c})$. If the laser frequency is lower than the resonance frequency it will be on the left side of the Doppler peak, but because of the Doppler shift the atom will see a frequency close to the Doppler peak and hence can have strong absorption. If the laser photon is moving in the same direction as the atom the atom will see a Doppler shifted frequency at $\nu\,(1 - \frac{v}{c})$. In such a case the frequency will be such that it will move away from the Doppler curve (Fig. 9.5) and hence the photon cannot be absorbed by the atom. If the laser beams are coming from both directions along an axis, both the beams can be absorbed by atoms moving in the opposite directions as the atoms in a gas are moving in all possible directions.

After absorption the atom will spontaneously emit a photon. In this process of emission, the atom will have a recoil momentum. This will be in a direction opposite to that of the direction of emission of the photon. This may cause an increase of momentum of the atom. But this is a random process. Hence the atom recoil will be in all directions and after several cycles, the net change due to spontaneous emission will average to zero. Hence after several cycles of absorption-spontaneous emission, the net change will be a reduction of velocity. In the entire process, the light beam exerts a force on the atom. This force may cause heating or cooling depending on the relative direction of atom motion and the laser beam propagation direction. In the following section we shall calculate the force acting on the atom due to incident light beam. In the subsequent section we shall consider the process of atom trapping.

9.3 Force acting on the atom in a laser field

If an atom absorbs a photon it will gain a momentum $\hbar\,\vec{k}$ because of the conservation of momentum. Similarly if an atom emits a photon it will acquire a momentum $-\hbar\,\vec{k}$. After an absorption the atom will decay to the ground state by spontaneous emission with the same frequency $\omega = k/c$. Hence after one cycle of absorption-spontaneous emission the net change of momentum of the atom is $\hbar\,(\vec{k} - \vec{k'})$, where $\vec{k'}$ is the wave vector of the spontaneously emitted photon in an arbitrary direction. If the process continues n times per second, after n cycles the net change of momentum is

$$\Delta\vec{p} = \hbar(n\,\vec{k} - \textstyle\sum_{j=1}^{n}\vec{k'}_j) = n\hbar\,\vec{k}. \tag{9.3.1}$$

Since the last sum will be zero when n is large, if the fluorescence rate is n, then the net change of momentum per unit time or force acting on the atom is

$$\vec{F} = n\hbar\,\vec{k}. \tag{9.3.2}$$

The rate of fluorescence is determined by the product of spontaneous decay rate from the upper level γ_b and the population of the upper level ρ_{bb} i.e. $n = \gamma_b\rho_{bb}$.

From the density matrix equations, described in Chapter 3, Eq. (3.3.8) and (3.3.9) we can write in the steady state

$$\rho_{bb} = -i\,\frac{\Omega}{2\gamma_b}\,(\tilde{\rho}_{ba} - \tilde{\rho}_{ab}) \tag{9.3.3}$$

and

$$\tilde{\rho}_{ba} = -i\,\frac{\Omega}{2}\,\frac{(2\rho_{bb}-1)}{(i\Delta+\gamma_{ab})}. \tag{9.3.4}$$

Combining the above two equations we get

$$\rho_{bb} = \gamma_{ab}\frac{\Omega^2}{2}\,\frac{1}{\gamma_b(\Delta^2+\gamma_{ab}^2)+\Omega^2\gamma_{ab}}. \tag{9.3.5}$$

Hence the force acting on the atom is

$$\vec{F} = \hbar\vec{k}\,\Gamma\,\frac{\Omega^2}{4\Delta^2+\Gamma^2+2\Omega^2}. \tag{9.3.6}$$

Where we have assumed the decay constants as

$$\gamma_b = \Gamma,\ \gamma_a = 0\ and\ \gamma_{ab} = \frac{\gamma_b+\gamma_a}{2} = \frac{\Gamma}{2}$$

This force, due to the radiation acting on the atom, is often called ponderomotive force.

The above results are valid for a stationary atom. For an atom moving with velocity $\vec{v}$, the atom will see a shifted frequency because of the Doppler shift. We can introduce an additional detuning due to the Doppler shift so that the expression for the ponderomotive force reduces to

$$\vec{F} = \hbar\vec{k}\,\Gamma\,\frac{\Omega^2}{4\left(\Delta+\vec{k}\cdot\vec{v}\right)^2+\Gamma^2+2\Omega^2} \tag{9.3.7}$$

In the case of low intensity of the laser beam, the Rabi frequency is small, so $\Omega^2 \ll \Gamma^2/2$. Hence we neglect the Rabi frequency term in the denominator. If the Doppler shift is small we can approximate

$$\frac{1}{4\left(\Delta+\vec{k}\cdot\vec{v}\right)^2+\Gamma^2} \cong \frac{1}{4\Delta^2+\Gamma^2+8\left(\vec{k}\cdot\vec{v}\right)\Delta}$$

$$\cong \frac{1}{4\Delta^2+\Gamma^2}\left(1-\frac{8\left(\vec{k}\cdot\vec{v}\right)\Delta}{4\Delta^2+\Gamma^2}\right).$$

Hence the force is

$$\vec{F} = \hbar\vec{k}\,\Gamma\frac{\Omega^2}{4\Delta^2+\Gamma^2} - \hbar\vec{k}\Gamma\Omega^2\,\frac{8\left(\vec{k}\cdot\vec{v}\right)\Delta}{(4\Delta^2+\Gamma^2)^2}. \tag{9.3.8}$$

We can write the above force in a simplified form as

$$\vec{F} = \vec{F}_0 - \hbar\vec{k}\Gamma M\alpha\left(\vec{k}\cdot\vec{v}\right), \tag{9.3.9}$$

where

$$\vec{F}_0 = \hbar\vec{k}\,\Gamma\frac{\Omega^2}{4\Delta^2+\Gamma^2}, \tag{9.3.10}$$

$$\alpha = \Omega^2\,\frac{8\,\Delta}{M(4\Delta^2+\Gamma^2)^2}. \tag{9.3.11}$$

$\vec{F}_0$ is the force due to the radiation acting on the stationary atom. When the atom is moving with a velocity, we have a velocity dependent force acting on the atom. If the atom and the radiation photon are moving in the same direction $\vec{k} \parallel \vec{v}$.

$$\vec{F} = \vec{F}_0 \pm \hbar \Gamma k^2 M \alpha \vec{v}. \tag{9.3.12}$$

The second term is a friction force. It is positive when the atom and photon are moving in opposite direction. This slows down the atom. If they are moving in the same direction, the force is negative and the atom speeds up. If the atom is located in a region with two opposite laser beams, the net force acting on the atom in the positive k direction is

$$\vec{F} = (\vec{F}_0 - \hbar \Gamma k^2 M \alpha \vec{v}) - (\vec{F}_0 + \hbar \Gamma k^2 M \alpha \vec{v})$$

$$= -2\hbar \Gamma k^2 M \alpha \vec{v} = -\beta \vec{v} \tag{9.3.13}$$

$$\beta = 2\hbar \Gamma k^2 M \alpha \tag{9.3.14}$$

This is the velocity dependent damping force acting on the atom. This leads to extremely low temperature in the micro-Kelvin region. Three counter propagating laser beams in three mutually perpendicular directions (Fig. 9.6) in space can effectively lower all three velocity components of atoms.

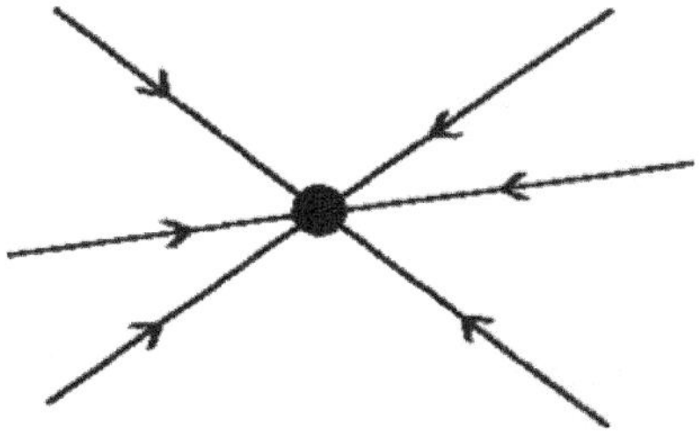

Fig. 9.6 Laser cooling by counter-propagating mutually perpendicular laser beams.

9.4 Magnetic Trapping

The first demonstration of the mechanical action of inhomogeneous magnetic fields on neutral atoms having magnetic moments is the Stern Gerlach experiment in 1924. This basic phenomenon was developed and extended for trapping of neutral atoms.

The cold atoms should be confined in a small region of space of the order a mm or less for short but finite duration of time in order to make an observation or an experiment on the cold atoms. This is the phenomena of trapping. Atoms can be trapped by static magnetic field or a combination of magnetic fields in multipolar, Helmholz or anti-Helmholz configuration or Joffe configuration. It is known that a magnetic field can produce shift or splitting of the energy levels of the atom. In inhomogeneous magnetic fields local energy minima are produced and the atoms can be confined in these local minima. For an atom at rest with a permanent magnetic moment $\vec{\mu}$ in a magnetic field $\vec{B}$, the energy shift is

$$\Delta E = -\vec{\mu}.\vec{B} = gJm\mu_B|B|. \tag{9.4.1}$$

The force produced is

$$\vec{F} = \vec{\nabla}(\vec{\mu}.\vec{B}), \tag{9.4.2}$$

where g is the Lande splitting factor, m is the projection magnetic quantum number and μ_B is the Bohr magneton. This field cannot produce a local energy maximum at rest in absence of a current, hence only static local energy minima are produced. The energy difference between the states is given by $\hbar\omega_L$, where $\omega_L = \mu B/\hbar$ is the Larmor frequency.

9.4.1 Anti-Helmholz Configuration

Helmholz configuration is a simple magnetic configuration produced by two parallel magnetic coils of diameter d separated by a distance $2l$, placed at $z = \pm l$ with the origin at the meeting point of the laser beams. A current of magnitude I flows in both the coils in the same direction. In anti-helmholz configuration the current flows in opposite directions in the parallel coils.

Two identical coils carrying opposite currents comprise a quadrupole trap (Fig. 9.7). This trap has a center where the field is zero. This is the simplest of all possible traps for the purpose of construction and optical access. This kind of trap was used for the first experiment on trapping of neutral laser cooled Rb atom at NIST, USA. The time duration of trapping was 1 sec and it was limited only by the background gas pressure.

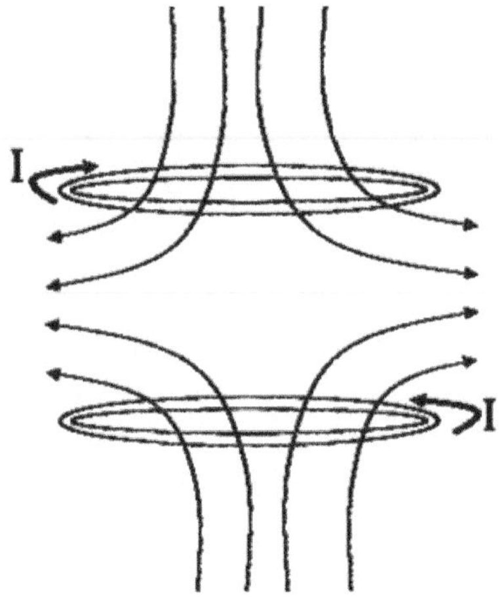

Fig. 9.7 Magnetic lines of force of a quadrupole trap in an anti-Helmholz configuration with two coils carrying electric currents in opposite directions. Magnetic field is maximum at the center of the two coils and is zero at central point on the axis passing through the centers of the two coils.

The magnetic field is zero at the center and increases in all directions as given by the expression

$$B = A\sqrt{\rho^2 + 4z^2}, \tag{9.4.3}$$

where $\rho^2 = x^2 + y^2$ and the field gradient has a constant value. Hence the force due to this field is not a harmonic force, neither is it a central force.

For circular orbits in the trap, we can set the magnetic force to be equal to the centripetal force

$$|\mu\,\nabla B| = Mv^2/\rho. \tag{9.4.4}$$

Hence the centripetal acceleration is $\vec{a} = \mu \vec{\nabla} B/\text{M}$, the angular frequency in the orbit of radius ρ in the z=0 plane is $\omega_c = \sqrt{a/\rho}$. If the trap size is of the order of a few cm and the magnetic field depth is nearly 100 Gauss, the acceleration can have a magnitude $a = 100\ \text{m}/sec^2$ and the fastest atom that can be trapped has a velocity $v = \sqrt{\rho a} \cong 1\ \text{m/sec}$.

9.5 Magneto-Optical Trap

9.5.1 MOT operation

We discussed in the earlier section how we can use six counter propagating laser beams in three mutually perpendicular directions to reduce all velocity components of atoms near the meeting point of all the lasers. If an anti-Helmholz configuration magnetic trap is produced with the coils carrying opposite currents situated on opposite sides at equal distance from the center of the optical confinement along any direction of the laser beam, say z, we have a combination of magnetic and optical fields. Such a hybrid configuration of optical confinement known as Magnet-Optical Trap (MOT). It was first successfully used in 1982 for trapping of neutral atoms. In this setup radiative cooling loads atoms in the trap. It depends on both the inhomogeneous magnetic field and the radiative selection rule in the presence of the polarized radiation beams. This kind of MOT is usually very robust in nature. The magnetic coils can be easily constructed in the laboratory and the atoms can be confined in a room temperature cell. The basic requirement is, of course, ultra-high vacuum. Normally one has to initially pump the cell to a high vacuum of 10^{-9} Torr and the pressure after the atoms are filled in should be of the order of 10^{-7} Torr. Very cheap low power diode lasers suitable for the alkali atom D-line transitions are available.

In a MOT slowly moving atoms are optically pumped and an inhogeneous magnetic field varying linarly with distance from the center i.e. $B(z) = Az$ is applied. For interpretation of MOT operation, we consider the simple case of $J = 0 \rightarrow J = 1$ transition. In a magnetic field, this will be split into three components $\Delta M = 0, \pm 1$. Because of the Zeeman shift the M =+1 level will be shifted up and the M= −1 level will be shifted down. The other level will remain unchanged. Since magnetic field varies with distance the field will be different at different distances from the center. Hence the Zeeman components will be tuned with the laser beam at different values of z Two counter propagating laser beams with opposite circular polarizations are detuned below the zero-field atomic resonance by Δ. As shown in Fig. 9.8 at a position z_1 the $\Delta M = +1$ transition is tuned away by the magnetic field, whereas the $\Delta M = -1$ transition is tuned closer to the resonance. Hence if the laser beam incident from the left side has σ^- polarization, more light will be scatterd by the beam. Similarly, if the laser radiation incident from the right side has a polariztion σ^+, more light will be scatterd by the beam. Both the beams will drive the atoms towards the center of the field, where the field is zero.

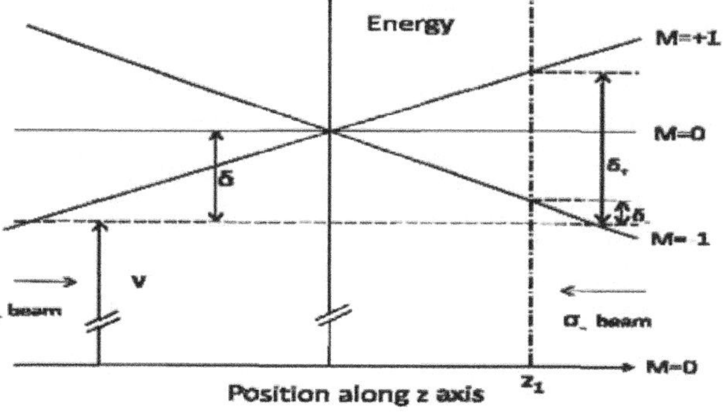

Fig. 9.8 One dimensional MOT operation.

In a MOT the force operates in the position space. This leads to simultaneous cooling and confinement in a MOT. As mentioned earlier, the MOT operation, presented for a one-dimensional case, can easily be extended to a three dimensional case when three laser beams are used and reflected by a mirror and passed through an appropriate polarizer. The above description is also valid for any atomic system with allowed transition $\Delta M = 0, \pm 1$. In the case of alkali atoms discussed in Chapter 4, we can use D type $S \rightarrow P$ transition for MOT operation. In common experiments, $5S_{1/2}(F = 3) \rightarrow 5P_{3/2}(F = 4)$ hyperfine component of Rb^{85} isoptope is used as the cooling transition..

9.5.2 Cooling and Trapping Force

From Eq. (9.3.6) the force acting on an atom can be written as

$$\vec{F} = \hbar\vec{k}\,\Gamma\,\frac{\Omega^2}{(2\Delta_\pm)^2 + \Gamma^2}. \tag{9.5.1}$$

Where we assumed Rabi frequency to be small so that it can be omitted in the denominator. The $\pm$ sign is used for two velocity directions of the laser beam as mentioned earlier. The detuning caused by the Doppler shift and the magnetic shift is

$$\Delta_\pm = \Delta \mp (\vec{k}\cdot\vec{v}) \pm \mu' B/\hbar, \tag{9.5.2}$$

where $\mu' = (g_b M_b - g_a M_a)\mu_B$ is the effective magnetic moment for the transition. When both the Doppler shift and Zeeman shift are small compared to the detuning Δ we can approximate the denominator following the same procedure as described in Sec. 9.3 and we obtain

$$\vec{F} = -\beta\vec{v} - \kappa\vec{r}, \tag{9.5.3}$$

where the spring constant, $\kappa = \mu' B/\hbar k$, arises from the Zeeman shift term. In the above, both the Doppler and Zeeman detuning terms have opposite signs for the counter-propagating beams. The above equation of the force is typically that of a damped simple harmonic oscillator with an oscillation frequency $\omega_c = \frac{\kappa^2}{M}$ and a damping rate β/M. The oscillation frequency is typically of the order of kHz if the magnetic field gradient is 10 Gauss/cm, the damping rate for a weak laser beam with Rabi frequency of the order of 10 MHz is usually 100 kHz. Thus the motion is overdamped. The restoring time to the center of the trap is a few msec for typical values of detuning and field gradient.

9.5.3 Experimental setup of the MOT

As shown in Fig. 9.9, the counter-propagating laser radiations with opposite circular polarizations (σ^+ and σ^-) are incident on the atomic cell from three mutually perpendicular directions. All the beams originate from the same laser and they are divided into three beams of equal intensities by the use of beam splitters. Each beam is reflected by a plane mirror after passage through the cell. This produces six counter-propagating beams in three directions in space. Quarter wave plates are used to produce the required circular polarization.

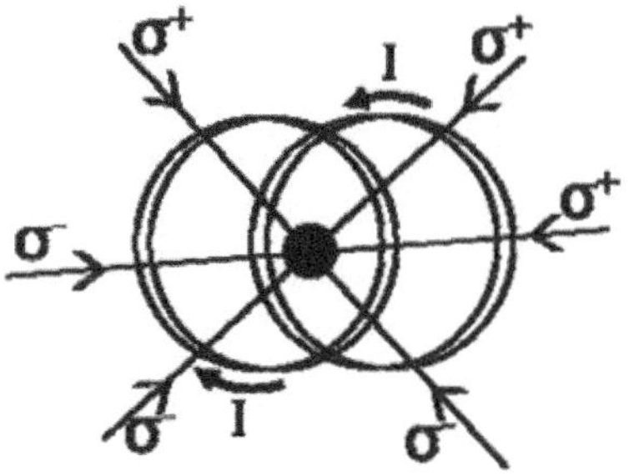

Fig. 9.9 Three dimensional cooling and trapping in an anti-Helmholz configuration. The central dark region represents the cooled and trapped atoms.

The pair of coils through which the electric current flows in opposite directions produces zero fields at the center. If the diameter of the coils is 5 cm and the separation of the coils is 7 cm and the number of turns in the coil is 80, a typical field gradient of 10 Gauss/cm is produced. This configuration leads to an efficient MOT.

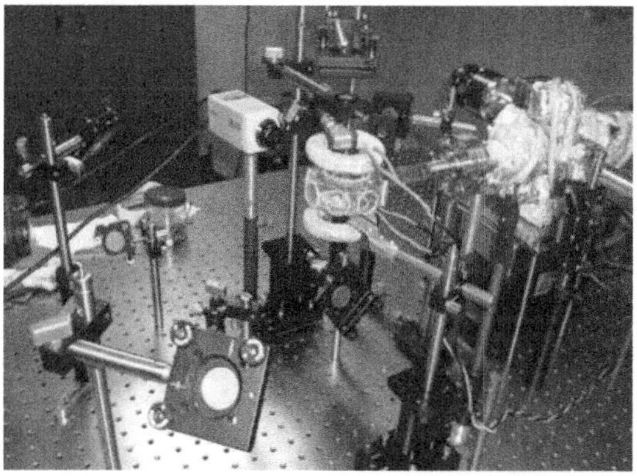

Fig. 9.10 Experimental setup for laser cooling and trapping of alkali atoms.

The Fig. 9.10 shows photograph of an experimental arrangement for the MOT that can cool and trap alkali atoms. A glass cell with ten windows (eight along the diameter and one at the top and one at the bottom of the cell) is used for confinement of atoms. The magnetic field coils are at the top and bottom of the cell. Six mutually perpendicular windows provide passage for the laser beams. Other windows are used for observation by CCD cameras and other experimental purposes with trapped atoms. The cell is connected to a getter valve for input of the atoms. The cell is also connected to a vacuum system provided by a combination of ion pump, a turbo and a backing pump. Normally diode laser beams with power of a few mW can be divided by a 30-70 and subsequently a 50-50 beam splitter to generate three beams of equal intensity.

Fig. 9.11 presents a sample photograph of the image of a cold cloud of Rb atoms taken by a CCD canera with number of pixels 512 x 512 and dimension of each pixel is 10 micron. The operating temperature is -25 C and the exposure time is 0.8 sec. The central white part is the picture of confined atoms. Other spots are reflected light from the cell. The shape of the cloud depends on the optical alignment and is very sensitive to any alteration of the experimental parameters.

Fig. 9.11 Image of a cold cloud of Rb atoms taken by a CCD camera.

9.5.4 Image analysis

The CCD records provide the photon counts against the number of pixels (Fig.9.12) in two dimensions of the image. A fit of the nearly Gaussisn spherical MOT image gives an estimate of the diametr of the cloud. In the above image, diameter of the atom cloud is nearly 0.4 mm.

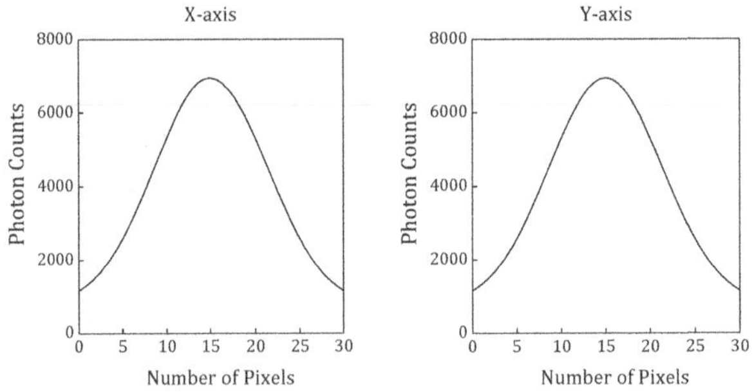

Fig. 9.12 Gaussian fit of the atom cloud in a MOT

The CCD records provide the photon counts against the number of pixels (Fig.9.12) in two dimensions of the image. The equipartition of the thermal energy over the degrees of freedom dictates that the energy is equal to the kinetic energy and the oscillator energy. Thus

$$k_B T = M v_{rms}^2 = \kappa r_{rms}^2. \tag{9.5.4}$$

From the value of diameter of the cloud and the typical parameters of the MOT as given here and the Problem 9.1 the temperature of the cloud can be estimated as 130 μK. The number of trapped atoms can be estimated from the intensity of the scattered light and is of the order of 10^6, so the number density of the atoms is 10^{10} $per\ c.c.$

9.6 Bose Einstein Condensation

9.6.1 Condensation of Boson

Bose's illustrious paper introducing a new statistitcs was published in 1924. It was translated into the German language and forwrded by Einstein for publication with a note that in English states the following:

"Bose's work signifies an important step forward. The method used here also yields the quantum theory of an ideal gas as I shall work out elsewhere".

The statement should be read carefully. Bose statistics was developed in the course of a new derivaion of Planck distribution law for radiation. It was applicable to *photons*. Einstein mentions that it can be applied to ideal gases as well and he promised that in future he would show that it could be used to develop the *quantum theory of an ideal gas*. He kept his words and published a paper the very next year and showed that gas phase atoms obeying Bose statistics can be cooled to very low temperature such that they will all condense to the ground state. *Dilute gas of atoms collected in the ground state will form a new phase.* The proposal of Einstein was seriously taken by scientists and efforts were made to understand the theoretical implications and to realize the new phase experimentally. After many unsuccessful attempts and some misinterpretaions, it was experimentally realized in 1995 after long seventy years.

We shall try to understand what this new phase is and also the problems for achiving the new phase.

9.6.2 Boson and Fermion

Bosons have integral spins, for example, photon, phonon or atoms like ^{87}Rb that have an even number of Fermions. On the other hand Fermions have half-integral spins like electrons, protons or ^{3}He or similar atoms. *Bosons allow any number of particles in a quantum state* ; so they are Gregarious in nature. In contrast, Fermi statistics allows only one particle in a state; so they are Loners in nature.

Table 9.1 Characteristics of Bose and Fermi particles

Bosons	Fermions
Integral spin	Half-integral spin
Examples: photons, Phonons, ^{87}Rb etc.	Examples: electrons, protons, ^{3}He etc.
Any number of particles in the ground state	Only one particle a state
Gregarious	Loners

9.6.3 Cold Atom as a de Broglie Wave

Cold atoms have low momentum p so the wavelength $\lambda_{dB} = h/p$ becomes large.

$$\lambda_{dB} = h\,/(2\pi mkT)^{1/2}\,.$$

Hence at low temperature, the average wavelength is much large and may become comparable to the average distance between the atoms (Fig. 9.13). This picture can be extended to the three dimensional case. The increase in deBroglie wavelength will mean that the probability distribution of the atom will be over a larger space or the atom itself will occupy more space. Since the distance between the atoms is fixed for a certain gas pressure, the

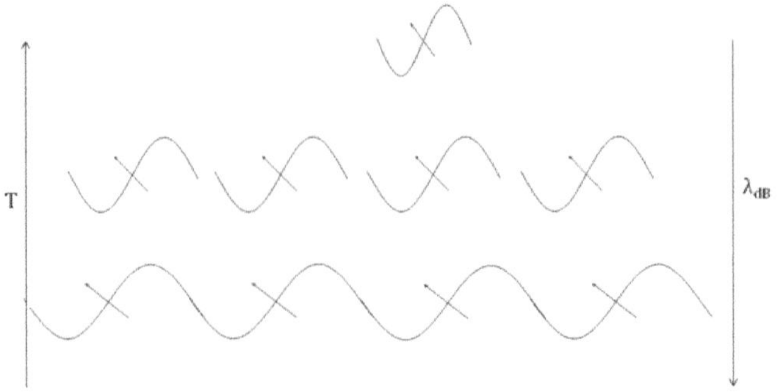

Fig. 9.13 Atom as a deBroglie wave. The wavelength increases with lowering of temperature. Finally, at very low temperatre one-dimensional BEC is formed.

average distance between the atoms decreases with lowering of temperature. Thus at a ceratin low temperature, there will be no space between the atoms. The waves will join each other. Hence there is a tight packing of the atoms in the coordinate space (Fig. 9.14). The dilute gas system becomes dense, as the atoms are "inflated".

At this low temperature and high density, a significant fraction of all the atoms will condense into a single ground state, since the Boltzman law states that the atoms will fall down to the lower state and Bose statistics allows accumulation of any number of atoms in a single state.

9.6.4 Quantum Identity Crisis

In this case, single atoms cannot be identified. What it really means is that the atoms are absolutely identical. There is no possible measurement that can separate them. Now you cannot separate one atom from another; they are all in the same place. Now we can see why it took so long before people could understand what BEC really meant. Atoms really can all be in the same place, though it goes against everything we see around us. It is only at the exceptionally low temperatures needed for BEC that they lose their individual quantum identities and coalesce into a single blob (Fig. 9.14). It may be called a "super atom" for that reason. All the atoms move in phase. They all dance in tune. Hence their dynamics cannot be described by a single Schroedinger equation. The system of all atoms that participate in the formation of the BEC is to be described by one equation of motion. Thus individual atoms are not described by one wave function. We have to develop the wave function describing the set of atoms that form BEC.

Fig. 9.14 Picture at the left shows atoms confined in a Bose Einstein Condensate. The picture on the right depicts a few cold atoms before the formation of BEC.

If we put an atom in a box, even a mixing bowl, it can only have certain particular energies. It has to choose from a particular set of allowed energies. What Einstein's equations predicted was that at normal temperatures the atoms would be in many different levels. However, at very low temperatures, a large fraction of the atoms would suddenly go crashing down into the very lowest energy level. *The atoms piling up in the bottom is what we call Bose-Einstein condensation.*

9.6.5 Theoretical aspects

Bose Einstein Statistics for a Perfect Quantum Gas

In a perfect quantum gas, the total Hamiltonian of a set of N non-interacting particles is

$$H = \sum_{i=1}^{N} h_i \,, \tag{9.6.1}$$

h_i are the one particle Hamiltonians. In terms of creation and annihilation operators, we can write the Hamiltonian and the number operator as

$$H = \sum_{\mu} e_{\mu} a_{\mu}^{\dagger} a_{\mu}, \qquad N_{op} = \sum_{\mu} a_{\mu}^{\dagger} a_{\mu}. \tag{9.6.2}$$

As discussed in Chapter 6, for a nultimode radiation field we can define the eigenstates in the Fock space as $|n_1, n_2, \dots \dots n_s \dots \rangle$.

For a given set of values of $n_1, n_2, \dots \dots n_s \dots$ we can use a notation j, then

$$H|j\rangle = E_j|j\rangle \,, \qquad E_j = \sum_s n_s \, e_s \tag{9.6.3}$$

$$N_{op}|j\rangle = N_j|j\rangle, \qquad N_j = \sum_s n_s \tag{9.6.4}$$

n_s is the occupation number of the individual quantum states and *it can have any positive integer value including zero.* The Bose Einstein distribution for the mean occupation number of the particles in the j-th state is

$$N_j = \frac{1}{e^{-(\mu - E_j)/k_B T} - 1} \,, \tag{9.6.5}$$

where μ is the chemical potential determined by the condition that the total number of particles is $N = \sum_j N_j$.

The Grand canonical partition function is

$$Z = \sum_j e^{(\mu - E_j) N_j / k_B T} \tag{9.6.6}$$

$$= \prod_s Z_s,$$

where,

$$Z_s = \sum_{n_s} e^{(\mu - e_s) n_s / k_B T} \tag{9.6.7}$$

is the grand canonical partition function for each single particle state s. This leads to a factoriztion of the partition function into a product of terms for individual quantum states.

For the Free energy, $\mathrm{F} = -k_B T \ln Z$, we can write

$$\mathrm{F} = \sum_s \mathrm{F}_s, \quad \mathrm{F}_s = -k_B T ln Z_s. \tag{9.6.8}$$

For Bosons the summation in Z leads to a geometrical series since n_s can have values 0,1,2,..... Hence

$$Z_s = \sum_{n_s} e^{(\mu - e_s)n_s/k_B T} = \frac{1}{1 - e^{(\mu - e_s)/k_B T}}. \tag{9.6.9}$$

The partition function satisfies the condition

$$lnZ = \sum_s ln \frac{1}{1 - e^{(\mu - e_s)/k_B T}}. \tag{9.6.10}$$

Hence we get

$$\mathrm{F} = k_B T \sum_s ln(1 - e^{(\mu - e_s)/k_B T}). \tag{9.611}$$

When we are dealing with a large system we have to know the number of particles N. For the chemical potential, we can write $\mu = -\left(\frac{\partial \mathrm{F}}{\partial N}\right)_{T,V}$.

Bose Einstein Condensation in a cubical box

For the non-interacting particles, we must have $\mu < 0$, this is necessary since the number of particles in a state should be positive. Considering E as the energy of the particles of mass M confined in a box of volume V= L^3 we can write the particle eigenfunctions and energies as

$$\psi(x, y, z) = \frac{1}{V} e^{i(p_x x + p_y y + p_z z)/\hbar}, \tag{9.6.12}$$

$$E = \frac{p^2}{2M} = \frac{p_x{}^2 + p_y{}^2 + p_z{}^2}{2M} \tag{9.6.13}$$

and the allowed values of momenta are

$$p_x = \frac{2\pi\hbar}{L} n_1, \ p_x = \frac{2\pi\hbar}{L} n_2, \ p_x = \frac{2\pi\hbar}{L} n_3, \tag{9.6.14}$$

where n_1, n_2, n_3 are integers. For a large box ($L \to \infty$), the momentum is quasi-continuous. In momentum space in a volume $dp_x dp_y dp_z$, the number of eigenstates is $V dp_x dp_y dp_z / (2\pi\hbar)^3$. Assuming continuous distribution of momentum the number of states with momentum below a certain value of $p = \sqrt{2ME}$ is

$$N(E) = \frac{V}{(2\pi\hbar)^3} \int_0^p 4\pi p^2 dp$$

$$= \frac{4\pi V}{3(2\pi\hbar)^3} (2ME)^{3/2}. \tag{9.6.15}$$

The density of states is

$$\rho(E) = \frac{dN(E)}{dE} = \frac{4\pi V}{(2\pi\hbar)^3} M^{3/2} (2E)^{1/2}. \tag{9.6.16}$$

It must be mentioned here that the number of states and also the density of states as calculated in this approximation is zero if the nergy is zero. Hence this density does not include the particles in the zero energy state.

The total number of particles is then

$$N = N_0 + \int_0^{\infty} dE\rho(E)\frac{1}{e^{-(\mu-E)/k_BT}-1}, \quad (9.6.17)$$

where N_0 is the number of particles in the ground or zero-energy state. Hence the chemical potential can be determined from the total number of particles. However, it has to be done numerically and analytical solution is complicated. For this purpose we define a few parameters and follow the standard nomenclature of statistical mechanics.

$$\beta = 1/k_BT, \qquad x = \beta E \quad (9.6.18)$$

and define fugacity as

$$z = e^{\beta\mu}. \quad (9.6.19)$$

Using the expression of density of states and the number of particles (Eq. (9.6.15) and (9.6.16)), we can write

$$\frac{N(2\pi\hbar)^3}{4\pi V}(M/\beta)^{-3/2} = \int_0^{\infty} dx\frac{\sqrt{2x}}{e^x/z-1} = I(z). \quad (9.6.20)$$

We have omitted the ground state number of particles in deriving this expression. The integral is a function of fugacity and hence the chemical potential. Since $\mu < 0$, we should have

$$0 \le z < 1 \quad (9.6.21)$$

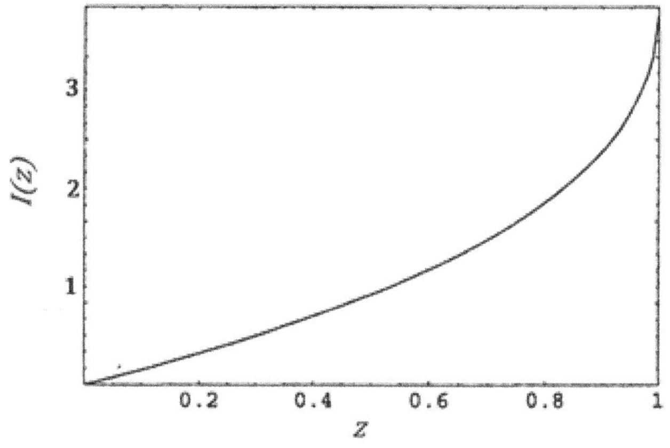

Fig. 9.15 Plot of $I(z)$ against z

In order to solve the above equation we can use a geometric method by plotting the curve (Fig.9.15) of the integral $I(z)$ against z and the intersection of a horizontal line for a fixed density and temperature with this curve will lead to a value of z. We need a value for the right hand side of the above equation when the horizontal line intersects the curve at $z < 1$. The integral has a value of approximately 3.274 when $z = 1$. Hence we need a solution when

$$\frac{N2\pi^2}{V}(M/\hbar^2\beta)^{-3/2} < I(1) \cong 3.274 \quad (9.6.22)$$

There is no solution, if the above is not satisfied. Hence for a fixed density $\frac{N}{V}$ we can define a critical temperature above which there can be a positive density for chemical potential satisfying the condition $\mu < 0$.

$$T_c = 3.308 \frac{\hbar^2}{k_B M}(\frac{N}{V})^{2/3}. \tag{9.6.23}$$

Thus the above condition demands that $T > T_c$. It may be noted that the above relation is valid for the case when the ground state population is zero. If we consider the ground state population we can find the situation that can be valid when $T < T_c$.

Accirding to Bose Einstein distribution

$$N_0 = \frac{1}{e^{-\beta\mu}-1}\ . \tag{9.6.24}$$

The chemical potential approaches the value zero from a negative value, hence the figacity approaches unity as $\mu \to 0$. Hence

$$N_0 = \frac{1}{e^{-\beta\mu}-1} = -\frac{1}{\beta\mu}. \tag{9.6.25}$$

Since $\beta\mu$ is a small negative quantity, the population of the ground level can be very large. If the energy of the first excited state is above the ground state by $k_B T$ or more,

$$N_1 = \frac{1}{e^{\beta(E_1-\mu)}-1} < 1. \tag{9.6.26}$$

The above is true because $\beta(E_1 - \mu)$ is a positive quantity with magnitude more than unity.

If the first excited state energy is less than $k_B T$, $\beta(E_1 - \mu) \ll 1$, so we can write

$$N_1 = \frac{1}{\beta(E_1-\mu)} = \frac{1}{\beta E_1 + 1/N_0}\ . \tag{9.6.27}$$

Hence,

$$\frac{N_1}{N_0} = \frac{1}{N_0\beta E_1+1} = \frac{\mu}{\mu - E_1} \cong -\frac{\mu}{E_1}. \tag{9.6.28}$$

Thus $N_1 \ll N_0$ as $\mu \to 0$. Hence the ground state population is indeed larger than any of the excited states. This leads to a transition to a state in which the population is mostly concentrated in the ground level at a temperature lower than the critical temperature. This exhibits *Bose Einstein Condensation* below the *critical temperature* $T_{c.}$.

Since $\mu = -\frac{1}{\beta N_0}$ and N_0 is very large below $T_{c.}$, the chemical potential is practically zero and hence the fugacity is nearly 1. Using the integral in the definition of the number of particles, N (Eq. 9.6.20) and also the expression of temperature (Eq. 9.6.23), it can easily be shown that

$$N_0 = N[1 - (T/T_{c.})^{\frac{3}{2}}] \tag{9.6.29}$$

This shows that as the temperature approaches the critical temperature the entire population collapses to the ground state.

9.6.6 Evaporative Cooling

We have noted in Section 8 that laser cooling in a MOT can cool the atoms down to a temperature of neraly 100 μK. This is much higher than the temperature needed for observation of BEC. As mentioned, the formation of BEC needs a low temperature and high density. The phase transition is predicted to occur at a phase space density of $\rho = n\lambda_{dB}^3$, where $n = \frac{N}{V}$. Using the expression of deBroglie wavelength we find from Eq. (9.6.22) that ρ should be of the order of unity. For real gases at NTP, this value is nearly 10^{-6}. The method of laser cooling also increases the phase space density, but it has a limit. The laser cooling is always limited by spontaneous emission and recoil energy. If the phase space density is too high, they start repelling each other; the light scattered by one atom is reabsorbed by another atom. The increase of phase space density also leads to higher rate of inelastic collisions involving a ground state and an excited state atom. This may lead to heating of the sample.

A more efficient method of atom cooling is Evaporative Cooling technique applied to atoms. It happens in the same principle as it happens in the cooling of a cup of hot tea (Fig.9.16). In the case hot tea at a temperature of 373 K, the molecules of tea, water etc have higher energy at the surface of the cup of tea than those at the lower part. When there is evaporation these molecules with higher kinetic energy and higher velocity will first leave the cup. The remaining molecules thermalize in a continuous process.

Thus the average energy of the cup of tea is lower leading to lopwering of temperature. This simple explantion in known from the school level physics. But the point to be noted here is that the number of molecules lost is very small. The loss can be easily measured by noting lowering of the height of the liquid tea in the cup that is allowed to cool by evaporation. When it cools down to room temperature (300K), there is decrease of temperature by more than 20%, though the number of molecules lost is not more than 2%. Hence the process is efficient.

Fig. 9.16 Evaporative cooling of a hot cup of tea.

The same physics of cooling a cup of tea is applied to cold atoms in a magnetic trap. The cold atoms in the MOT are driven to a magnetic trap produced by a magnetic field. The atoms with magnetc moments align themselves parallel to the direction of the field and the magnetic field interaction energy is $-\vec{\mu}.\vec{H}$, where $\vec{\mu}$ is the magnetic moment of the atom and $\vec{H}$ is the magnetic field as shown in the Figure (Fig. 9.17). The interaction is attaractive and produces a potential well trapping the atoms, if the atomic moments and the magnetic field are in the same direction. Atoms within the trap have different velocities and the higher velocity or higher energy atoms are near the top of the trap. If the atoms have a magnetic moment in the reverse direction the interaction energy will be $+\vec{\mu}.\vec{H}$ and hence it produces a repulsive interaction. The potential curve will be reversed as shown in Fig. 9.17. Thus if a mechanism can invert the moment of the atoms they will escape the well.

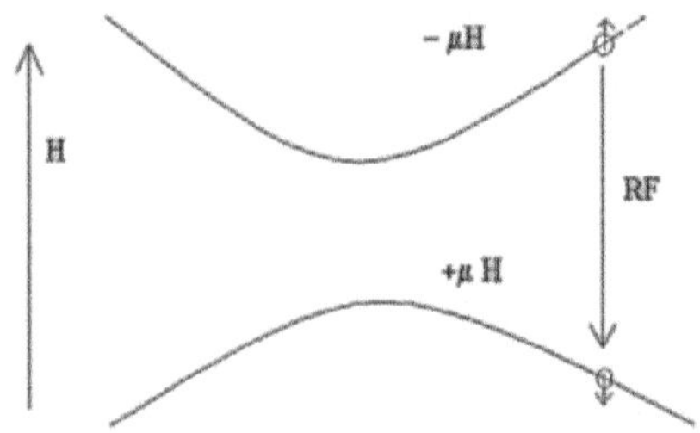

Fig. 9.17 A radio frequency field is applied to reverse the spin of the atoms that move from an attractive to a repulsive potential well.

This is achieved by using a radio frequency field with a frequency ν_{RF} such that $h\nu_{RF} = 2\vec{\mu}.\vec{H}$. This frequency can be tuned to coincide with some particular trap depth. If the frequency is set to a value that corresponds to the top of the well, then the atoms near the surface of the potential well will move out of the top as they have a reversal of the moment with gain of energy.

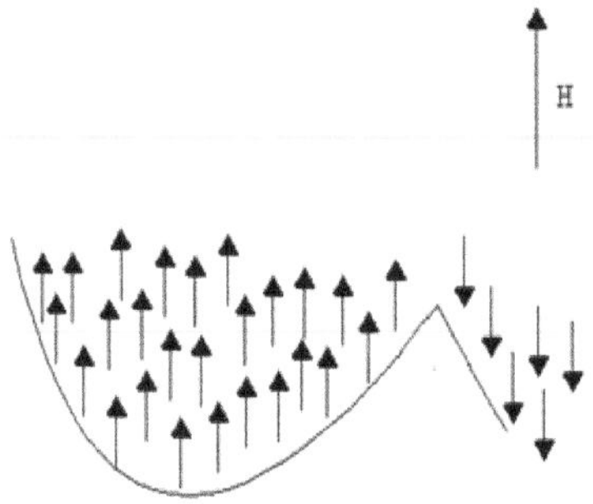

Fig. 9.18 Evaporative cooling of atoms in a magnetic bowl by driving out atoms with higher velocity

This process (Fig.9.18) may be contiued and the RF frequency can be tuned and slowly lowered to remove more atoms from the top of the trap. In this process, all the atoms can be removed if we reach the bottom of the trap. In practice, the tuning of the frequency is stopped at a certain depth of the well. The well will contain the atoms that will retharmalize, after atoms with higher velocity have been driven out, and will have lower average velocity. In this way, temperature close to nano-Kelvin could be reached.

9.6.7 Observation and application of BEC

The observation of BEC is mostly destructive. Any laser beam incident on the BEC may evaporate out the atoms and completely destroy the condensed state.

In the destructive method of observation, we can only get the velocity distribution of the atoms giving a signature that the BEC was formed. In this method as soon as the cooling lasers are switched off, the cold atoms start flying out with their residual velocities. They are recorded by a CCD camera kept fixed near the BEC cell after a few msec from the moment of switching off the lasers. The atoms with larger initial velocity will traverse a longer path and the slower atoms will travel a shorter distance. So a velocity distribution can be observed in the coordinate space. This will give a signature of BEC(Fig. 9.19). Before the formation of BEC ($T>T_c$) the atomic velocity distribution can be described by a Gaussian desribing the cold atoms. As the condensation begins ($T \approx T_c$) we can see a much narrower and sharper peak sitting on a Gaussian pedestal. The sharp peak describes the BEC atoms and the broader background represents the atoms that have not yet been condensed. With further lowering of temperature ($T<T_c$) we can get a single narrow Gaussian describing all the condensed atoms. The condensation was observed for Na atom at MIT and Rb atom at NIST, colorado. The picture recorded by camera for the velocity distribution resembles the curves shown in Fig. 9.19. When all the atoms form BEC it is difficult to measure temperature as the velocity distribution curve is very sharp. As mentioned at the beginning of this section when all atoms form BEC they all move in phase and it is not possible to have a thermal distribution.

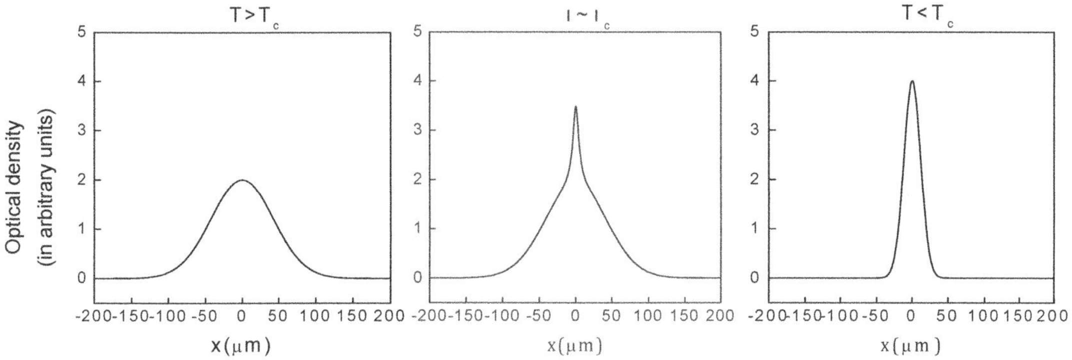

Fig. 9.19 The velocity distribution of atoms recorded as a function of distance traversed measured from the source after a fixed time interval of a few msec from the moment of switching off the cooling laser.

9.6.8 Applications

Observation of BEC has been a breakthrough in physics and opened the door for an immaculately large number of theoretical and experimental work. Theoretical work is a challenge to the existing theories of many-body physics.

The major interest is to explore how the atoms stay together in a small region in the condensed form, although there is no chemical bonding between the atoms. The atom-atom interaction is very weak and mean field thery is applicable. For theoretical studies, people normally use Gross-Pitaevsky equations, that are basically non-linear Schroedinger equations developed in the sixties for calculation of properties of nuclear matter. Thermodynamic properties are also investigated in detail.

A large number of technical applications have also been envisaged and attempted. BEC atoms have been used as a coherent beam of atoms to develop atom chips of micrometric size. Since the atom beams are coherent in nature any beam of atoms from a Bose Einstein condensate can act like an atom laser having high energy that will have very strong penetrating power because of the coherent character of the beam. Many other applications are also considered.

Problems

9.1 Calculate the rms velocity of atoms of atomic mass 100 at normal temperature and pressure.

9.2 An atom and a photon moving along a line collide with each other and the photon is absorbed. If the laser wavemgth is 795 nm and the atomic mass is 87, Find the cahnge in velocity in a single atom-photon collision.

9.3 If the Doppler shift of the photon frequency as seen by an atom moving with velocity 100 m/sec. is 50 MHz, find the radiation wavelength.

9.4 If the Doppler shift is large, obtain an expression of the ponderomotive force by retaining term upto the next higher power of velocity in Eq. (9.3.13).

9.5 Calculate the ponderomotive force if the saturation broadening is high.

9.6 Consider an anti-Helmholz configuration with the coils separated by a distance of 14 cm and the diameter of the coils are 6 cm. If the magnetic field at the centre of the coils is 40 Gauss, find the field gradient and the magnitude of the field near the coils.

9.7 If the angular frequency in an orbit of radius of 4 cm in the z=0 plane is 50 Hz, find the acceleration of the atoms and the velocity of the fastest atoms. Calculate the magnitude of the magnetic force on Rb (isotope 85) atoms in the trap.

9.8 Consider a MOT with Rb (isotope 85) with the laser detuned by 24 MHz from the D_1 transition with an effective magnetic moment 6.83 x 10^{-24} erg/Gauss, magnetic field gradient 11.6 Gauss/cm, the decay constant 24 MHz and the Rabi frquency 19.5 MHz, Calculate the harmonic oscillation frequency of the MOT.

9.9 Calculate the deBroglie wavelength of Rubidium (isotope 85) atoms at temperatures 300K, 100 μK and 200 nK.

9.10 If the critical temperature for transition to Rb BE Condensate is 20 nK, find the number density of atoms in the condensate.

9.11 Calculate the phase space density of real gases at N. T. P. Calculate the same for the condensate at temperature 100 nK and pressure of 2 nano-Torr.

Further reading

1. P. Meystre, Atom Optics, Springer, New York, 2001.
2. H. L. Metcalf and P. Van der Straten, Springer, New York, 1999
3. J.T. Mendonc¸a and H. Terc¸as, *Physics of Ultra-Cold Matter*, Springer Series on Atomic, Optical, and Plasma Physics , Springer , 2013.
4. P. Meystre and M. Sargent III, Elements of Quantum Optics, Springer, Berlin, 2007.
5. M. O. Scully and M.S. Zubairy, Quantum Optics, Cambridge University Press, Cambridge, 1997.
6. S. C. Rand, Non-linear and Quantum Optics, Oxford, New York, 2010.
7. C. Cohen-Tanoudji, Laser Cooling and Trapping of Neutral Atoms-Theory, Phys. Reports, 219, 153, 1992.
8. V.I.Balykin and V.S. Letokhov, Atom Optics with Laser Light, Harood Academic Publishers, Chur, 1995.
9. D. Kleppner, Bose Einstein Condensation, Physics Today, 49, 13, 1996.
10. W. D. Phillips, Laser Cooled and Trapped Atoms, Progress of Quantum Electronics, 8, 115, 1984.
11. .M. Ueda, Fundamentals and New Frontiers of Bose Einstein Condensation, World Scientific, Singapore, 2010.
12. C.J. Pethick and H. Smith, Bose Einstein Condensation in Dilute Gases, Cambridge University Press, Bristol, 2010.
13. A. Griffin, D. W. Snoke and S. Stringari, Bose Einstein Condensation, Cambridge University Press, Bristol, 2010.

Chapter 10
Lasers in Quantum Information Science

10.1 Quantum entanglement

10.1.1 EPR Paradox

Even after successful analysis of experiments carried out so far, there are questions on basic interpretation of wave function that forms the core of quantum mechanics. The orthodox view of quantum mechanics is that after the measurement the system collapses to a wave function that gives the result of measurement of the physical property. But according to the realistic view of Einstein, Podolsky and Rosen, the system has the property even before the measurement since it gives a certain result in the measurement. Hence according to EPR paradox quantum mechanics is an incomplete theory, since there is some external information that is necessary for interpretation of physical reality. Although an agnostic or metaphysical view is to ignore the conceptual foundation of the theory.

A simple idea of EPR paradox can be presented in terms of the dissociation of a diatomic molecule in the singlet state (S=0) into two atoms having opposite spins. After dissociation the two atoms will move in opposite directions. If the atom 1 has spin up with angular momentum $+\hbar/2$, then the other atom should have spin down with angular momentum $-\hbar/2$, since the molecule initially had a zero spin angular momentum. It can also be the other way round, i.e. the atom1 may have spin down and atom 2 can them have spin up. The atomic state is then

$$|\psi\rangle = \frac{1}{\sqrt{2}} \left[\, |1 \uparrow\rangle |2 \downarrow\rangle - |1 \downarrow\rangle |2 \uparrow\rangle \right] \qquad (10.1.1)$$

Quantum mechanics tells us that both results are possible with fifty per cent probability for each. But it tells that if one atom is in the spin up state, it is definite that the other atom is in the spin down state. Thus the results of measurement are correlated. After the dissociation the atoms may move away from each other by thousands of kilometers. When a measurement on one atom gives a certain result (say, spin up) then another measurement thousands of kilometers away must give the other result (i.e. spin down). EPR paradox says that the two atoms already had a spin arrangement and this is a reality and quantum mechanics did not know it. Orthodox view of collapse of wave function was called by Einstein "spooky action at a distance" and this view cannot be true. The basic assumption of EPR paradox is that no information can propagate faster than light. If the information on collapse of the wave function travels to the other location after a finite time, then any instantaneous measurement of the spin of the second atom would lead to equal probability of finding it with either spin up or spin down. This will lead to a probability of finding both atoms having the same spin. This will violate the principle of conservation of angular momentum.

10.1.2 Bell's inequality

According to the EPR paradox, the wave function does not describe the entire picture, we need some other variable, called hidden variable λ, in order to describe a state. It is a hidden variable because we have no way to describe or measure it. Bell considered a modification of EPR experiment in which detectors were not parallel to each other, but they could be rotated independently. We consider that one of them is directed along a direction $\vec{A}$ and the other along a direction $\vec{B}$; they can measure spin in units of $\hbar/2$. If the detector oriented at $\vec{A}$ measures the spin +1, then the detector oriented at $\vec{B}$ measures the spin -1. Bell proposed to measure the average of the product of spins registered for different configurations of the detectors and he called the average as $J(\vec{A},\vec{B})$. As an example if the detectors are parallel this product is always -1, hence $J(\vec{A},\vec{A}) = -1$; but if they are anti-parallel, we get $J(\vec{A},-\vec{A}) = +1$. For arbitrary orientations we get from quantum mechanics

$$J(\vec{A},\vec{B}) = -\vec{A}.\vec{B} \tag{10.1.2}$$

Now we consider that the result of measurement from each dissociation event depends on the hidden variable λ, we do not know its mathematical form. The detector at B can be oriented by its observer in an arbitrary manner just before the atom reaches the detector at A so that the observer at A does not know it. This is the locality condition since the information at the end B will take time to reach A. We can define a function $\alpha(\vec{A},\lambda)$ for the measurement at A and also define function $\beta(\vec{B},\lambda)$ for the meaustrement at B. They can both have values ± 1. For parallel alignment of the detectors we can have perfect correlation for each λ

$$a(\vec{A},\lambda) = -\beta(\vec{A},\lambda) \tag{10.1.3}$$

The average of the product of measurements for different distributions λ may be obtained as

$$J(\vec{A},\vec{B}) = \int \rho(\lambda)\ a(\vec{A},\lambda)\beta(\vec{B},\lambda)d\lambda, \tag{10.1.4}$$

$\rho(\lambda)$ is the probability distribution for the variable λ. So it is positive and is normalized to unity, or $\int \rho(\lambda)\,d\lambda = 1$. From Eq. (10.1.4) we can write

$$J(\vec{A},\vec{B}) = -\int \rho(\lambda)\ a(\vec{A},\lambda)\alpha(\vec{B},\lambda)d\lambda. \tag{10.1.5}$$

If $\vec{C}$ is another unit vector then

$$J(\vec{A},\vec{B}) - J(\vec{A},\vec{C})$$

$$= -\int \rho(\lambda)\,[\,a(\vec{A},\lambda)\alpha(\vec{B},\lambda) - a(\vec{A},\lambda)\alpha(\vec{C},\lambda)]d\lambda$$

$$= -\int \rho(\lambda)\,[1 - a(\vec{B},\lambda)\alpha(\vec{C},\lambda)]a(\vec{A},\lambda)\alpha(\vec{B},\lambda)d\lambda. \tag{10.1.6}$$

Since $|a(\vec{B},\lambda)|^2 = 1$.

But we know that

$$|a(\vec{A},\lambda)\alpha(\vec{B},\lambda)|=1 \tag{10.1.7}$$

and

$$\rho(\lambda)[1 - a(\vec{B},\lambda)\alpha(\vec{C},\lambda)] \geq 0. \tag{10.1.8}$$

Hence

$$|J(\vec{A},\vec{B}) - J(\vec{A},\vec{C})| \leq 1 + J(\vec{B},\vec{C}) \tag{10.1.9}$$

This is Bell's inequality. This is true for any local hidden variable theory without any assumption on the nature and number of hidden variables. It should be noted here that this does not hold true for quantum mechanics as stated in Eq. 10.1.2. For any two vectors $\vec{A}$ and $\vec{B}$ perpendicular to each other and $\vec{C}$ a vector in an arbitrary direction in the same plane containing the other two vectors, $J(\vec{A},\vec{B}) = 0$ and $J(\vec{A},\vec{C})$ and $J(\vec{B},\vec{C})$ are negative and less than one. Hence $|J(\vec{A},\vec{C})|$ cannot always be less than or equal to $1 + J(\vec{B},\vec{C})$.

An experiment on two photon atomic correlation with randomly placed detectors showed that quantum mechanics gives correct result and Bell's inequality is not compatible. But Bell's result holds only for local hidden variables and does not apply for non-local hidden variables. However, nature can be non-local.

10.2 Quantum teleportation

An important application of quantum entanglement is quantum teleportation or quantum information transfer. In this process, a quantum state is transferred from one place to another that may be wide apart. There is no particle transfer in the process. We shall define quantum bits or qubits, instead of classical bits of traditional classical computers.

10.2.1 Qubits

Any two level quantum system can form a qubit. Examples of such systems are two hyperfine levels of an atom or a combination of the ground state and an excited state of an atom or two orthogonal states of polarization of a single photon. The hyperfine qubits have a very long life time that can span for years. The optical level qubits also have a relatively long life time of the order of second. Logic gate operation time in a computer is of the order of microseconds. In quantum information science the qubit levels are usually denoted as $|0\rangle$ and $|1\rangle$. As an example for an atom the ground state $|a\rangle$ corresponds to $|0\rangle$ and the excited state $|b\rangle$ corresponds to $|1\rangle$.

Bell States

We define Bell States as maximally entangled two-qubit states.

$$|\Phi_1\rangle = \frac{1}{\sqrt{2}}[\,|00\rangle + |11\rangle], \tag{10.2.1}$$

$$|\Phi_2\rangle = \frac{1}{\sqrt{2}}[\,|00\rangle - |11\rangle], \tag{10.2.2}$$

$$|\Psi_1\rangle = \frac{1}{\sqrt{2}}[\,|01\rangle + |10\rangle], \tag{10.2.3}$$

$$|\Psi_2\rangle = \frac{1}{\sqrt{2}}[\,|01\rangle - |10\rangle]. \tag{10.2.4}$$

These states are orthogonal and normalized and they form a useful basis for description of any 2-qubit system in quantum teleportation.

10.2.2 Procedure

The teleportation procedure is popularly described by an experiment by two partners Alice and Bob. Alice wants to transfer quantum information to her friend Bob located at an unknown distant place. She attempts it by choosing a state of a qubit in the form

$$|\psi\rangle = a|0\rangle_A + b|1\rangle_A \tag{10.2.5}$$

where a and b are arbitrary complex numbers. The suffix A denotes that it belongs to Alice. For transfer of information she uses a maximally entangled state or the Bell state. For an example we consider the Bell state $|\Phi_2\rangle$ as defined in Eq. (10.2.2)

$$|\Phi_2\rangle = \frac{1}{\sqrt{2}}[\,|0\rangle_A|0\rangle_B - |1\rangle_A|1\rangle_B]. \tag{10.2.6}$$

In order to be explicit we have designated the qubits by the subscripts to indicate the states that belong to Alice and Bob. In this procedure in the presence of the unknown state to be teleported, the total state of the system is $|\psi\rangle \oplus |\Phi_2\rangle$. From the above equations we can write

$$|\psi\rangle \oplus |\Phi_2\rangle = \frac{1}{\sqrt{2}}[\,a|0\rangle_A|0\rangle_B|0\rangle_A - b|0\rangle_A|0\rangle_B|1\rangle_A - a|1\rangle_A|1\rangle_B|0\rangle_A + b|1\rangle_A|1\rangle_B|1\rangle_A]$$

$$= 1/2[|\Phi_1\rangle_A(a|0\rangle_B - b|1\rangle_B) + |\Phi_2\rangle_A(a|0\rangle_B + b|1\rangle_B) - |\Psi_1\rangle_A(a|1\rangle_B - b|0\rangle_B) - |\Psi_2\rangle_A(a|1\rangle_B + b|0\rangle_B)]. \tag{10.2.7}$$

The total state is now written in terms of the Bell states that belong to Alice. If Alice makes a measurement she will know her state, at that instant the result can project Bob's system to the corresponding state. As an example, if Alice finds from the measurement that her Bell state is $|\Phi_1\rangle_A$, then the state for Bob will be projected to the state $(a|0\rangle_B - b|1\rangle_B)$. Now Alice needs to transfer the information of the result of her measurement to Bob, so that he can make a suitable unitary transformation to know the state of his system. So Alice has to use a classical communication channel to send the information of the result of her measurement. This process needs a finite time, since information transfer cannot be faster than light.

10.3 Quantum Computers

10.3.1 Qubits in Quantum Computation

In traditional computers information is coded in terms of bits. The first computers were realized with vacuum tubes in the middle of 20-the century. By the latter half of the century information processing could be done much faster by using semiconductor transistors and with the development of integrated circuits the computers could be miniaturized. In the 21-st century attempts are being made to replace bits by qubits. In the case of bits we have a binary system consisting of 0 and 1. A classical three bit computer can only store 8 numbers like 000, 001, 010, 011, 100 101, 110 and 111. In a qubit system we can have linear combinations of all the 8 qubit states as

$$|\varphi\rangle = c_1|000\rangle + c_2|001\rangle + c_3|010\rangle + c_4|011\rangle + c_5|100\rangle + c_6|101\rangle + c_7|110\rangle + c_8|111\rangle.$$

Such a choice of quantum superposition of states gives enough flexibility to a quantum computer. A quantum computer can be in a coherent superposition of all 2^N N-bit numbers. A quantum operation is a unitary transformation that simultaneously operates on all the 2^N states, and not on a single state or qubit as the classical computers do. This offers the possibility of parallel computation.

Experimental realization of quantum circuits, algorithms and information processing is very challenging. We can discuss some of the guiding principles on systems based optical photons from lasers.

10.3.2 Physical realization of Quantum Computer

General requirements

Qubits must be isolated, but accessible. This may seem to be contradictory, but a delicate balance is required. Two time scales are important: (i) time the system remains quantum mechanically coherent and (ii) the time required to perform a unitary operation. These time scales are related and depends on how strongly the system is coupled to the external effects.

Capability of preparing the initial states is an important issue. Examples are creation of single photon states or cool ions to their ground state.

Means of performing the universal family of unitary transformations must be known. Unitary transformations can be realized from single spin qubit operation and controlled-NOT gates. Experimental quantum computation aims to achieve these gates. Unitary transformations should not lead to decoherence.

Method of measurement of the output result should also be known. As an example, a qubit state $a|0\rangle + b|1\rangle$ of a two-level atom may be measured by pumping to the excited state and measuring the fluorescence. If the photomultiplier detects the signal, the qubit would collapse to the state $|1\rangle$ with probability $|b|^2$. Otherwise it would collapse to state $|0\rangle$ with probability $|a|^2$

Optical Photons

Optical photons constitute suitable physical system to represent a qubit. They are zero charge particles and do not strongly interact with each other and also with most other matters. They can be guided over long distance through optical fibers with low loss. They exhibit quantum phenomena like double slit interference. They can interact with each other in non-linear optical media.

As we discussed in Chapter 6, energy of simple harmonic waves can be quantized in units of $\hbar v$ that represents the energy of a single photon. A cavity can have a superposition of two states, a zero photon and a one photon state thus forming a qubit $a|0\rangle + b|1\rangle$ as introduced in Eq. (10.2.5) for an arbitrary state. We can consider two cavities and consider a state as a combination of two states as being in one cavity (10) or the other(01). We shall consider single photon states. Such states can be obtained from the laser beam by attenuating the laser output. The laser output can be expressed as a coherent state and as defined in Chapter 6 (Eq. 6.9.21), this is a combination of n-photon states and is an eigen state of the photon annihilation operator.

$$|\alpha\rangle = e^{-|\alpha|^2/2} \sum_n \frac{\alpha^n}{\sqrt{n!}} |n\rangle.$$

It can be easily shown that $\langle n|\alpha|n\rangle = |\alpha|^2$. By attenuating the laser beam we can generate a weaker coherent state and can have only one photon. As an example, if $|\alpha| = 0.1$, $|\alpha\rangle \cong \sqrt{0.995}|0\rangle + \sqrt{0.099}|1\rangle + \ldots\ldots$ This is thus a state when 99.5% of the time it is in zero photon state or no photon comes at all. These sources cannot be synchronized; it is difficult to know when a photon comes out or not. Better synchronization can be obtained by using non-linear media such as KH_2PO_4. It can generate two photons that will conserve energy and momentum. The states can be synchronously detected. Unfortunately, the best known non-linear Kerr media are weak and cannot provide an efficient cross phase modulation between single photon states. Hence realization of quantum computers based on optical photons is difficult.

Ion Trap

In an ion trap quantum computer qubit is formed with the hyperfine levels of the atom or ion (internal energy of the atom) and vibrational energy or phonons arising from external motion of the ions in a trap. For initial state preparation laser cooling and trapping method (Chapter 9) is used to cool the ions to the lowest vibrational or zero phonon modes. This process aims to reduce the temperature that reflects the kinetic energy of the ions such that $k_B T \ll \hbar\omega_z$, where ω_z is the oscillation frequency of the ion in the trap along the z-axis., The potential for ion motion in the trap is actually anharmonic for larger displacements. Width of the ion oscillation in the trap should be small compared to the wavelength of the incident laser radiation. We want to study individual ions with the laser beam.

Ions can be prepared in a qubit state by the method of optical pumping described in Chapter 4. A laser radiation applied to the ions can couple one of the lower states to another excited state. Eventually the ion will decay from this state to another lower state that is not coupled by the laser with any other state. So it will stay there. This state has a long life time. If the ion decays back to the initial state, it will again be excited and the atoms may be transferred to the other state. For unitary evolution, arbitrary transforms by laser pulses are used to externally manipulate the atomic states. The qubits interact via shared phonon states. For readout hyperfine level populations are measured.

The major drawback of ion trap as quantum computer is the short life time of the phonon states. But the life time of the internal energy states (hyperfine) are much longer. We can capitalize on that. A string of 40 mercury ions have been trapped along the axis of the trap. Such systems may someday be successfully used as quantum computer.

It may be noted here that physical realization of a quantum computer is not feasible now. But there are encouraging experiments and it is hoped that it may soon be practicable in a limited sense.

Problems

10.1 Show that

$$|\psi\rangle \oplus |\Phi_1\rangle = \frac{1}{\sqrt{2}}[\, a|0\rangle_A|0\rangle_B|0\rangle_A + b|0\rangle_A|0\rangle_B|1\rangle_A + a|1\rangle_A|1\rangle_B|0\rangle_A + b|1\rangle_A|1\rangle_B|1\rangle_A]$$

$$= 1/2[|\Phi_1\rangle_A(a|0\rangle_B + b|1\rangle_B) + |\Phi_2\rangle_A(a|0\rangle_B - b|1\rangle_B) + |\Psi_1\rangle_A(a|1\rangle_B + b|0\rangle_B) + |\Psi_2\rangle_A(a|1\rangle_B - b|0\rangle_B)]$$

Where $|\psi\rangle$ is an arbitrary qubit state belonging to Alice as defined in Eq. (10.2.5) and $|\Phi_1\rangle$ is the maximally Entangled state or Bell state for A and B defined in Eq. (10.2.1)

10.2 The value of nuclear magneton is $5\,x\,10^{-27}$ Joules/ Tesla. Compare the energy of the nuclear spin at a magnetic field of 8 Tesla with the thermal energy at a temperature 300 K.

10.3 If $|\alpha| = \sqrt{0.1}$ what is the probability of a zero photon state ?

Further Reading

1. D. J. Griffiths, Introductory Quantum Mechanics, Prentice-Hall, New Jersey, 1994.
2. C. C. Gerry and P. L. Knight, Introductory Quantum
3. Optics, Cambridge University Press, Cambridge, 2005.
4. P. Meystre and M. Sargent III, Elements of Quantum Optics, Springer, Berlin, 2007.
5. M. O. Scully and M.S. Zubairy, Quantum Optics, Cambridge University Press, Cambridge, 1997.
6. M. A. Nielsen and I. A. Chuang, Quantum Computation and quantum Information, Cambridge University Press, Cambridge, 2005.
7. J. P. Martikainen, Physical Realization of a Quantum Computer, http://www.helsinki.fi/~jamartik.
8. J. S. Bell, Speakable and Unspeakable in Quantum Mechanics, Cambridge University Press, Cambridge, 1987.
9. Einstein, B. Podolsky, and N. Rosen. Can quantum-mechanical description of physical reality be considered complete? Phys. Rev., 47:777–780, 1935.

Index

Printed and bound by CPI Group (UK) Ltd, Croydon, CR0 4YY
17/10/2024
01775694-0001